Palgrave Studies in Audio-Visual Culture

Series Editor
K.J. Donnelly
School of Humanities
University of Southampton
Southampton, UK

The aesthetic union of sound and image has become a cultural dominant. A junction for aesthetics, technology and theorisation, film's relationship with music remains the crucial nexus point of two of the most popular arts and richest cultural industries. Arguably, the most interesting area of culture is the interface of audio and video aspects, and that film is the flagship cultural industry remains the fount and crucible of both industrial developments and critical ideas.

Palgrave Studies in Audio-Visual Culture has an agenda-setting aspiration. By acknowledging that radical technological changes allow for rethinking existing relationships, as well as existing histories and the efficacy of conventional theories, it provides a platform for innovative scholarship pertaining to the audio-visual. While film is the keystone of the audio visual continuum, the series aims to address blind spots such as video game sound, soundscapes and sound ecology, sound psychology, art installations, sound art, mobile telephony and stealth remote viewing cultures.

More information about this series at
http://www.palgrave.com/gp/series/14647

Milena Droumeva • Randolph Jordan
Editors

Sound, Media, Ecology

palgrave
macmillan

Editors
Milena Droumeva
Simon Fraser University
Burnaby, BC, Canada

Randolph Jordan
Concordia University
Montréal, QC, Canada

Palgrave Studies in Audio-Visual Culture
ISBN 978-3-030-16568-0 ISBN 978-3-030-16569-7 (eBook)
https://doi.org/10.1007/978-3-030-16569-7

Cover illustration: Gary James Joynes Photo Title: 12 Tones Edition: 434.97 Hz
Cover Design: eStudio Calamar

This Palgrave Macmillan imprint is published by the registered company Springer Nature Switzerland AG
The registered company address is: Gewerbestrasse 11, 6330 Cham, Switzerland

Acknowledgements

This volume owes its existence to the field of acoustic ecology founded by R. Murray Schafer at Simon Fraser University, and we are grateful for the invaluable opportunity to study and work with the legacy of the World Soundscape Project in various forms. As part of this legacy, we would like to acknowledge the incredible mentorship and long-time friendship of WSP pioneers Barry Truax and Hildegard Westerkamp; the collegiality of the Acoustic Crew working group led by Barry in the Sonic Research Studio, where Milena and Randolph met in 2012; the inspiring transdisciplinarity of this volume's contributors, some of whom were also members of the Acoustic Crew over the years; and the instrumental feedback by the collection's various reviewers, especially Neil Verma, whose own work on sound media has been an inspiration to us both throughout this process. We would also like to thank Lina Aboujieb for her enthusiasm in embracing this project at Palgrave, series editor Kevin Donnelly for believing in the value of this project right away, and Ellie Freedman for her diligence in overseeing the production of the book. We thank Stacey Copeland for her assistance in preparing the manuscript for submission and artist Gary James Joynes for permission to use one of his wonderful "frequency paintings" for our cover (a more appropriate image would be hard to imagine). And last but not least, the editors acknowledge each other's unwavering support and collegiality throughout this process and our respective families, who inevitably bore the brunt of our unavailability during the crunch times.

Historical Timeline for the Development of Acoustic Ecology at Simon Fraser University Through to the Foundation of the World Forum for Acoustic Ecology

1965

Founding of Simon Fraser University (SFU) in Burnaby, British Columbia. Composer R. Murray Schafer arrives that same year and founds the Electronic Music Studio in the basement of the SFU Theatre within the Centre for Communication and the Arts. Here Schafer begins his work on music and soundscape education that leads to the development of the World Soundscape Project (WSP) and the Sonic Research Studio within a few years' time, laying the foundations for the field of acoustic ecology.

1967–1969

Schafer publishes his first book, *Ear Cleaning: Notes for an Experimental Music Course* (1967), which presents a set of exercises used in a first-year music course at SFU. He follows quickly with *The New Soundscape* (1969), detailing his observations about the rapidly changing urban soundscapes of the post-war era.

1970

Birth of the World Soundscape Project with Schafer's publication of *The Book of Noise* (1970), written as an educational primer on urban noise pollution based largely on his experience of rapid changes in the Greater Vancouver area that encompasses Burnaby and SFU. *The Book of Noise* is retroactively labelled as Document No. 1 in the World Soundscape Project publication series, though this designation does not appear on the 1970 edition published by Price Milburn Inc.

1971

Work is completed on *Okeanos*, a 90-minute quadrophonic tape composition produced by members of the WSP, consisting of ocean recordings with related sonic treatments and spoken word content. *Okeanos* is labelled as Document No. 2 in the World Soundscape Project publication series. This is arguably the earliest example of soundscape composition as a genre of electroacoustic music composition embedded within the research and media practices of acoustic ecology.

1972

SFU forms its new Department of Communication Studies and Schafer moves the Electronic Music Studio into new facilities there, renaming it the Sonic Research Studio and designating it as the official home of the World Soundscape Project. With financial support received from the Donner Canadian Foundation, Schafer begins research and study of the sound environment.

The WSP publishes *A Survey of Community Noise By-laws in Canada*, a compendium of all existing noise legislation in Canada, with suggestions for improvements. This is labelled as Document No. 4 in the World Soundscape Project series (despite being dated one year before Schafer's essay *The Music of the Environment*, labelled as Document No. 3).

In September, WSP members Howard Broomfield, Bruce Davis, and Peter Huse begin making recordings around Vancouver, continuing until November 1976. Most of the recordings are made over the first year and provide the material included on *The Vancouver Soundscape* release in 1973.

1973

Hildegard Westerkamp joins the World Soundscape Project after contacting R. Murray Schafer about her experience using *The Book of Noise* in her classroom as a music teacher in Vancouver.

Barry Truax joins the World Soundscape Project after an invitation from R. Murray Schafer based on recognition of his skills as an electronic music composer and sound engineer studying under Otto Laske at the Institute of Sonology, Utrecht University.

Schafer publishes his essay *The Music of the Environment* to outline his ideas of "acoustic design" as an alternative approach to the negative connotations of noise abatement campaigns. The essay is published by UNESCO in both the English and French versions of the first issue of its *Cultures I* journal. Schafer would also publish the essay as "No. 1 of an Occasional Journal devoted to Soundscape Studies" to be edited by Schafer himself. This would develop into a book series entitled *The Music of the Environment* dedicated to publishing the research of the World Soundscape Project. This series first overlaps with, and then takes over from, the World Soundscape Project document series. *The Music of the Environment* essay is listed as Document No. 3 in the WSP document series, and is retroactively labelled as Document No. 1 in *The Music of the Environment Series.*

The World Soundscape Project releases *The Vancouver Soundscape* as a double-LP and companion book presenting research and recordings into the present-day sounds of the City of Vancouver and its historical development over the past century. Together the book and LP are listed as Document No. 5 of the World Soundscape Project, and would retroactively be designated as Document No. 2 in *The Music of the Environment Series.* The booklet is originally self-published by the WSP, and reprinted in 1978 by ARC Publishers along with the later WSP publications. WSP members credited with collecting and writing the book materials include Howard Broomfield, Bruce Davis, Peter Huse, Colin Miles, and R. Murray Schafer. Howard Broomfield, Bruce Davis and Peter Huse record the tapes, and these are edited for release by Bruce Davis. R. Murray Schafer edits the written texts. Research assistants listed include Kathleen Swink, Barry Truax, Betty Anne Wang, and Hildegard Westerkamp.

From September to November, WSP members Bruce Davis and Peter Huse set out on a cross-Canada trip to make recordings of a variety of different sound environments reflecting Canada's sprawling geography.

1974

A special issue of the journal *Sound Heritage* (Vol. 3, No. 4) is published featuring essays and interviews with members of the WSP, including Hildegard Westerkamp's influential "Soundwalking" article, which continues to be used as the basis for conducting and teaching the art of soundwalking today.

In October, the CBC Radio programme *Ideas* airs the ten-part *Soundscapes of Canada* series based on the cross-Canada recordings made by Bruce Davis and Peter Huse the previous year. Each programme is written and edited by different members of the WSP, working solo and in pairs, including Howard Broomfield, Bruce Davis, Peter Huse, R. Murray Schafer, and Barry Truax.

1975

From February to June a larger contingent of the WSP team heads to Europe to make recordings across a variety of different sound environments that would lead to the publication of the books *European Sound Diary* (1977) and *Five Village Soundscapes* (1978). This cohort includes Howard Broomfield, Bruce Davis, Peter Huse, R. Jean Reed, and R. Murray Schafer. Hildegard Westerkamp also works on this project informally.

Schafer resigns his post at SFU to move east and focus on his writing and career as a composer. The Sonic Research Studio and the World Soundscape Project are handed over to Barry Truax, who remains director of both until his retirement in 2015, at which time he creates the Glenfraser Endowed Professorship in Sound Studies to carry on the work, and Milena Droumeva is hired into the position.

1977

The WSP issues two publications documenting their trip to Europe in 1975: *European Sound Diary*, a narrative account of their travels, and *Five Village Soundscapes*, presenting the research they conducted there. These are labelled as Document No. 3 and Document No. 4 in *The Music of the Environment Series* respectively. The latter book is reprinted in 2009 as part of a new Finnish study entitled *Acoustic Environments in Change* that revisits the same five villages and adds a sixth case study from Finland to the mix.

Schafer publishes *The Tuning of the World* at McClelland & Stewart in Canada and Alfred A. Knopf in the U.S.A., a summation of his ten years of research with the World Soundscape Project at SFU. The book is reprinted as *The Soundscape: Our Sonic Environment and the Tuning of the World* in 1994 by Destiny Books. This book is widely regarded as the foundational text for acoustic ecology.

1978

The WSP publishes *Handbook for Acoustic Ecology*, edited by Barry Truax. The book is labelled as Document No. 5 in *The Music of the Environment Series*, and is the last entry in that series. This also stands as the last book to be published by the WSP and the Sonic Research Studio. Barry Truax publishes a new edition of the handbook in CD-Rom format in 1999 through his Cambridge Street Records label founded in 1985.

Hildegard Westerkamp begins a weekly radio programme on Vancouver Co-operative Radio entitled *Soundwalking*, exploring the educational and expressive potential of attention to environmental sound. The programme only runs through 1979, but Vancouver Co-op Radio continues its commitment to related content through programmes like *Soundscape*, currently led by the Cut and Run artist collective, which includes participation from Brady Marks and Helena Krobath, both trained in acoustic ecology at SFU.

1984

Barry Truax publishes his book *Acoustic Communication* formalizing his communicational approach to the core concepts developed by the World Soundscape Project for the study of sonic environments. A second edition is published in 2001 by Ablex Press. This book is used as a textbook for undergraduate courses in acoustic ecology in the School of Communication at Simon Fraser University for the next 25 years.

1985

Barry Truax founds Cambridge Street Records for specialist releases in computer and electroacoustic music, beginning with his own *Sequence of Earlier Heaven*, composed specifically for the LP format.

1988

Hildegard Westerkamp completes her MA in Communication at SFU. Her thesis, supervised by Barry Truax, is titled *Listening and Soundmaking: A Study of Music-as-Environment*. She begins teaching courses in acoustic communication at SFU while pursuing her graduate studies and continues until 1990.

Around the same time, in 1986, composer Susan Frykberg also comes to SFU for graduate study with Barry Truax. She develops an online version of an undergraduate course in acoustic ecology and authors a textbook for it, remaining to teach it for the next seven years.

1989

Hildegard Westerkamp releases her composition *Kits Beach Soundwalk*, which becomes a benchmark in soundscape composition as an artistic offshoot of acoustic ecology.

1990

Digitization of the analogue reel-to-reel tapes from *The Vancouver Soundscape*, *Soundscapes of Canada*, and *Five Village Soundscapes* projects begins, with Susan Frykberg transferring them onto the new Digital Audio Tape (DAT) format.

1991

The Soundscape Newsletter is founded by the World Soundscape Project. They publish six issues through mid-1993 with Hildegard Westerkamp as the main editor. The goal of the newsletter is to facilitate an international network of people working in and around the field of acoustic ecology.

WSP member Robert MacNevin begins a new set of Vancouver recordings using the DAT format that would run from November 1991 through June 1995. This set of tapes provides the material for the *Soundscape Vancouver 1996* release.

1993

The World Forum for Acoustic Ecology (WFAE) is founded during the Banff International Conference on Acoustic Ecology, *The Tuning of the World* (Aug. 8–14). This formalizes acoustic ecology as a field now practised internationally, and the WFAE takes over publication of *The Soundscape Newsletter*. Hildegard Westerkamp remains the principle editor until the final print edition of the newsletter in December 1995. The newsletter would eventually be reborn online as part of the WFAE website, managed by retired acoustic ecology educator Gary Ferrington. The

WFAE also founds *Soundscape: The Journal of Acoustic Ecology*, published in print from 2000 to 2015 and is subsequently issued as a digital publication, now in its 17th volume.

1997

The World Soundscape Project at SFU comes full circle with a re-issue of their 1973 release of *The Vancouver Soundscape*, now on CD for the first time at Cambridge Street Records. This version contains a few alterations from the original release (notably the loss of a side-long conversation between Schafer and the sound recordists, a substantial remix by Barry Truax of the track entitled *The Music of Various City Quarters*, and the addition of Hildegard Westerkamp's recording of *New Year's Eve in Vancouver Harbour 1980–81*). This re-issue is paired with a second CD entitled *Soundscape Vancouver 1996* that includes new recordings made in the early 1990s by Robert MacNevin and an audio documentary by Hildegard Westerkamp and Barry Truax exploring changes to the Vancouver soundscape across 20 years. The release also features a series of soundscape compositions by a variety of international members of the WFAE that creatively engage the WSP archive across those same 20 years, including Sabine Breitsameter, Darren Copeland, Claude Schryer, and Hans Ulrich Werner. This marks the WSP's first attempt at a longitudinal study of a single geographical locale documented in their archive. The release caps the four-week *Soundscape Vancouver 1996* composition workshop organized by the WFAE (May 6–June 8, 1996), culminating with a final concert on June 7 as part of the International New Music Festival, and a one-day symposium on June 8 co-organized by the School of Communication at SFU, the WFAE, and the Goethe Institute Vancouver. *Soundscape Vancouver 1996* provides an early example of the new international collaborations fostered by the WFAE and marks a shift from the emphasis on Vancouver as the home of acoustic ecology to a truly international field of research and practice.

Coda

A third set of recordings is produced in 2010–2011 via a small grant from the Social Sciences and Humanities Research Council (SSHRC) obtained by Jan Marontate, faculty at the School of Communication at SFU. Recordists for this set include Vincent Andrisani, Nathan Clarkson, and subsequently Jenni Schine, but as yet no follow-up publication has been issued to engage with this later material. Around this time, Nathan Clarkson creates a new set of digital transfers of all the original WSP tapes using the latest in digital capture software, and the archive is placed online and made available to numerous researchers around the world. Currently, under the direction of Milena Droumeva with the assistance of Helena Krobath, the 1970s archive is being organized within SFU Library's digital repository (RADAR) with updated metatags, to be made available on an open-source basis starting in 2019.

World Soundscape Project Publications

World Soundscape Project Document Series:

Document No. 1: *The Book of Noise* (1970)
Document No. 2: *Okeanos* (1971)
Document No. 3: *The Music of the Environment* (1973)
Document No. 4: *A Survey of Community Noise By-laws in Canada* (1972)
Document No. 5: *The Vancouver Soundscape* (1973)

The Music of the Environment Series:

Document No. 1: *The Music of the Environment* (1973)
Document No. 2: *The Vancouver Soundscape* (1973)
Document No. 3: *European Sound Diary* (1977)
Document No. 4: *Five Village Soundscapes* (1977)
Document No. 5: *Handbook for Acoustic Ecology* (1978)

Contents

Notes on Contributors

Mitchell Akiyama is a Toronto, Canada, based scholar, composer, and artist, and an assistant professor at the Daniel Faculty of Architecture, Landscape, and Design at the University of Toronto. His eclectic body of work includes writings about plants, animals, cities, and sound art; scores for film and dance; and objects and installations that trouble received ideas about history, perception, and sensory experience. Akiyama's output has appeared in commensurately miscellaneous sources such as *Leonardo Music Journal*, *ISEA*, *Sonar Music Festival*, *Raster-Noton Records*, *Gendai Gallery*, and in many other exhibitions, publications, and festivals.

Vincent Andrisani, PhD, is a term lecturer in Simon Fraser University's School of Communication in Vancouver, BC. Intersecting the fields of sound studies, sensory ethnography, and urban studies, his research uses the everyday sounds of the city to explore urban transformation and development. www.vincentandrisani.com

Leah Barclay is the director of Biosphere Soundscapes and a research fellow at Griffith University in Australia, where she is leading a portfolio of research in acoustic ecology and climate change. She specializes in electroacoustic music, acoustic ecology and emerging fields of biology exploring environmental patterns and changes through sound. Her work has been commissioned, performed and exhibited internationally by organizations including UNESCO, Ear to the Earth, Al Gore's Climate Reality and the IUCN. Barclay is the President of the Australia Forum for Acoustic Ecology and the Vice President of the World Forum for Acoustic Ecology.

Sarah Barns is a research fellow based at the Institute for Culture and Society, Western Sydney University, Australia, who works across sound and creative media through her digital placemaking practice, Esem Projects. Barns was awarded a three-year Urban Studies Postdoctoral Research Fellowship by the Urban Studies Foundation for the project 'Platform urbanism: the role of city labs, data infomediaries and open government experiments in urban governance'. Barns's research builds on her doctoral thesis to locate and critique the geographical imaginations of contemporary digital urbanism and situate a creative sound-led response in the city of Sydney.

Karin Bijsterveld is a historian and professor of Science, Technology and Modern Culture at Maastricht University in the Netherlands. She is the author of *Mechanical Sound: Technology, Culture, and Public Problems of Noise in the Twentieth Century* (2008), *Sound and Safe: A History of Listening behind the Wheel* (2014, with E. Cleophas, S. Krebs & G. Mom), co-editor of *Sound Souvenirs: Audio Technologies, Memory and Cultural Practices* (2009, with J. van Dijck), and co-editor of the *Oxford Handbook of Sound Studies* (2012 with T. Pinch). She has recently published *Sonic Skills* (Palgrave, 2019), a book available in open access.

Milena Droumeva is Assistant Professor of Communication and Glen Fraser Professor in Sound Studies at Simon Fraser University specializing in mobile technologies, sound studies, and multimodal ethnography. Droumeva has a background in acoustic ecology and works across the fields of urban soundscape research, sonification for public engagement, as well as gender and sound in video games.

Mikael Fernström (Softday) is the Director of the Interaction Design Centre at University of Limerick (UL), Ireland. He has several backgrounds—in electronics, telecoms, computer science, music and interaction design. He joined UL in 1996 and founded the Interaction Design Centre with Prof. Liam Bannon. He is the principal investigator on several research projects ranging from the EU IST Future and Emerging Technology Sounding Object (2001–2012) to Enterprise Ireland sponsored MetaMusic (2006–2008), to HEA PRTLI sponsored by FutureComm (2007–2011). As an artist, he has collaborated under the name Softday since 1999 on a range of art-science projects from Bliain Le Baisteach (2000) to Amhran na Beacha (2014).

Özgün Eylül İşcen is a PhD candidate in Computational Media, Arts, and Cultures at Duke University, United States. She has a background in film and soundscape studies with a focus on intercultural cinema and experimental documentary. She works on media theory and history in the context of the Middle East. Her dissertation project situates digital media in the Middle East in wider flows of capital, technology, and culture, thus offering a deeper understanding of the material infrastructures and ideological underpinnings of contemporary media. She received her BA in Psychology and Sociology from Koç University, Turkey, and MA in Interactive Arts and Technology from Simon Fraser University, Canada.

Randolph Jordan is Visiting Assistant Professor of Cinema at Concordia University in Montréal, Canada. His research, teaching, and creative practice reside at the intersections of soundscape research, media studies, and critical geography. He has published widely on the ways in which the fields of acoustic ecology and film sound studies can inform each other.

Andra McCartney is Professor Emerita of Communication Studies at Concordia University, Canada. She is a soundwalk and radio artist and has published articles with *Music and Arts in Action*, *Organised Sound*, *Leonardo Music Journal*, *Perspectives of New Music*, *Musicworks*, *Resources for Feminist Research*, and *Borderlines*, as well as chapters in edited volumes, including *Performing Nature*, *Aural Cultures*, *Gender and Music*, and *Ghosts in the Machine*. Her research project "Soundwalking Interactions" investigates the ways that people listen to and engage with soundwalks and artworks made from soundwalks.

Linda O Keeffe is the head of the School of Fine Art at Edinburgh College of Art, Edinburgh University. She has exhibited and performed in Asia, the USA, Canada and Europe. Recent works include an installation at EV&A, the Irish Biennale, in 2016, a sound design commission for the Hidden Theatre project with the 42nd Street organisation in Manchester in 2017, the premier of a new performance work at Cibelo in Madrid in 2017, and a UK touring exhibition, Hybrid Soundscape I-IV for the *Sounds Like Her* exhibition during 2017–2020. Her written publications include 'The Sound Wars: Silencing the Working Class Soundscape of Smithfield' published in Politiques de Communication in 2018. She is the founder of the Women in Sound Women on Sound organisation, www.wiswos.com, and editor in chief of the *Interference Journal*, www.interferencejournal.org

Jonathan Sterne is James McGill Professor of Culture and Technology at McGill University, Canada. He is the author of *MP3: The Meaning of a Format* (2012), *The Audible Past: Cultural Origins of Sound Reproduction* (2003), and of numerous articles on media, technologies and the politics of culture. He is also the editor of *The Sound Studies Reader* (2012). His projects consider instruments and signal processing; artificial intelligence; and the intersections of disability, technology and perception. His is writing *Diminished Faculties: A Political Phenomenology of Impairment* and, with coauthor Mara Mills, *Tuning Time: Histories of Sound and Speed.* http://sterneworks.org

Sean Taylor is an Ireland-based artist and researcher. He is completing a PhD in Acouscenic Listening and the Creative Soundwalk at Limerick Institute of Technology. He is programme joint-leader in the Social Practice and the Creative Environment MA and lectures on the Sculpture & Combined Media programme at Limerick School of Art & Design. He holds an MA in Fine Art from the University of Ullster in Belfast and a postgraduate scholarship from the Kunstenacademie, Rotterdam, Holland. Taylor established the Art & Science collaboration *Softday* with software engineer Mikael Fernström in 1999.

Barry Truax is a composer and a professor emeritus in the School of Communication (and formerly the School for the Contemporary Arts) at Simon Fraser University, Canada. He worked with the World Soundscape Project, editing its *Handbook for Acoustic Ecology* (1978) and producing the book *Acoustic Communication* (1984/2001). As a composer, his multi-channel soundscape compositions are frequently featured in concerts and festivals around the world. Since his retirement in 2015, he has been the Edgard Varèse Guest Professor at the Technical University in Berlin and a guest composer at the 2016 BEAST Festival in Birmingham. www.sfu.ca/~truax

Heikki Uimonen worked as a professor of popular music research at the Sibelius Academy, University of the Arts, Helsinki, in 2012–2017. He is an ethnomusicologist and a docent on acoustic communication and soundscape studies at the Universities of Tampere and Eastern Finland. He has published articles, a monograph, and edited anthologies on music consumption, radio music, compact cassettes, and changing sonic environments. His research interests include the sonic construction of place, mediated music, social use of music, and transforming soundscapes. He is the board member of Finnish Society for Acoustic Ecology.

Hildegard Westerkamp is a British Columbia, Canada, composer, lecturer, founding member of the World Forum for Acoustic Ecology (WFAE) and co-editor of the *Soundscape* journal. She has lectured and conducted soundscape workshops internationally. Her compositional work focuses on the sounds of the environment and, most importantly, draws attention to the act of listening itself. Her written work has been published in journals and books such as *Organized Sound*, *Radio Rethink*, *Kunstforum*, *Musicworks*, *MusikTexte*, and *Sounding Art*. She was a researcher for the *World Soundscape* and taught acoustic communication at Simon Fraser University. http://www.hildegardwesterkamp.ca

List of Figures

CHAPTER 1

Sound, Media, Ecology: Introduction in Three Acts

Milena Droumeva and Randolph Jordan

Acoustic ecology made its international debut in 1977 with *The Tuning of the World*, the most popular and widely cited publication to emerge from the research carried out by the World Soundscape Project (WSP) under the direction of composer R. Murray Schafer at Simon Fraser University (SFU) in Vancouver, British Columbia, Canada. Acoustic ecology is an interdisciplinary framework for documenting, analysing, and transforming sonic environments: an early model of the cross-boundary thinking and multi-modal practice now common across the digital humanities. Yet, in the past 40 years, acoustic ecology has not prompted the global rethinking of architectural design or urban planning that the field prescribed, and instead, questions about its biases and research methods have dominated academic conversations. With the emergence of sound studies and the burgeoning expansion of "ecological" thinking across a wide variety of disciplines in recent years, much of what interested early acoustic ecology

M. Droumeva (✉)
Simon Fraser University, Burnaby, BC, Canada
e-mail: mvdroume@sfu.ca

R. Jordan (✉)
Concordia University, Montréal, QC, Canada
e-mail: randolph.jordan@concordia.ca

M. Droumeva, R. Jordan (eds.), *Sound, Media, Ecology*,
Palgrave Studies in Audio-Visual Culture,
https://doi.org/10.1007/978-3-030-16569-7_1

has been taken up in differing contexts at a wilful distance from the problematic constructions of the original WSP. This volume will serve as a comprehensive investigation into the ways in which current scholars working with sound are re-inventing acoustic ecology across diverse fields, drawing on acoustic ecology's focus on sensory experience, place and practice-based research, and attendance to social structures and politics. From sounding out the Anthropocene across diverse cultural contexts, to rethinking our auditory media landscapes, this volume brings acoustic ecology's central problems to questions around space, cultures, technology, and mediated sensorialities.

Contributors include two leading figures from the original generation of the WSP, Barry Truax and Hildegard Westerkamp, along with leading scholars in sound studies today like Jonathan Sterne and Karin Bijsterveld and a host of other key figures contributing to the discourse. We open with a section on theorizing the field where prominent voices in contemporary sound studies problematize and explicate rich aspects of acoustic ecology histories, drawing germane conversations to the fore. The volume then moves on to consider the ways in which environment and community have been investigated in and through sound, bringing together a variety of settings and perspectives. The third section is focused on culture and society, including mediation, technological and otherwise, in the study of sound. Together, this collection demonstrates the livelihood of ecological work that engages with sound today, allowing the original concerns of acoustic ecology to dialogue with a wide range of work being done in critical sound studies, radical cultural history, science and technology studies, media theory, eco-humanities, and the digital arts.

As editors of this volume, we are using this introduction to chart pathways through the intersections of sound, media, and ecology along three lines: the historicity of ecological thinking across disciplines, acoustic resonances of the Anthropocene, and the spaces of new sound pedagogies. What follows is a telematic dialogue (as we work on the collection from the West and East coasts of Canada) about themes and ideas that have most inspired our thinking about where this volume fits in the contemporary landscape of sound-based research and practice across art, media, and ethnography. Above all, we grapple with one of the central tensions at the very heart of ecology: understanding how ecological systems work requires a sense of totalizing knowledge at a distance, while the prevailing wisdom of ecological thinking argues that we can only ever understand an ecosystem from inside, as part of the system. This has been a long-standing source of tension in the field of acoustic ecology. For example, the implied mastery of Schafer's goal of comprehensive acoustic design stands in marked contrast to the participatory engagement of Hildegard Westerkamp's prescription for soundwalking and sound art as a mode of "speaking from inside the soundscape" (Westerkamp, 2001). The tension between these macroscopic and microscopic perspectives can be charted across the WSP's impulse towards exhaustive documentation, sonic cartography, historical contextualization, and all the things that make ecological study itself a difficult balancing act. Needless to say, the spread of ecological thinking into disciplines and fields outside the natural sciences has created multiple permutations of how we might do research and create art along ecological lines. Here are our thoughts about how these struggles are made manifest within the pages of this volume as a reflection upon the state of soundscape research today.

Act 1. Everything New Is Old Again

Milena:

I was listening to a podcast recently—as so many stimulating ideas seem to come from these days—in which the host was interviewing someone about a utopian community based on the principles of something called "social permaculture." Permaculture, if you haven't heard, is the fashionable new term for urban farming, or small-scale vegetable/food production for oneself. People from North America go to expensive workshops all the way to Bali and Indonesia to study permaculture: the art of working with the soil and natural conditions, a sort of harmony of elements working together to produce sustainable long-term food supply. The podcast was drawing on that term to articulate utopian community relations as a kind of social permaculture: a society (a commune really) that functions in tune with the environment, in balance with nature and with respect and fairness to the reproduction of equitable social order. A society that presumably has transcended what bell hooks calls "capitalist white supremacist patriarchal order" (2000) in favour of liberation for all in a permanent, sustainable way.

What struck me about the idea of "social permaculture" was its remarkable resemblance to the "ecological" thinking of the 1960s and 1970s that brought us acoustic ecology in the first place. An anti-modernist return-to-the-human-scale of things, acoustic ecology has always seemed a utopian term, an attempt to transpose arguably romanticized notions about the natural-ness of nature into fraught social relations, to avert tensions by striving for balance. Acoustic ecology offers a proposal for reclaiming humanistic values and has long been a movement to *quiet* an increasingly loud, diverse, challenging, technologized world.

Fast forward several decades, utopian social permaculture is similarly an attempt to navigate an impending climate crisis, as well as salvage humanity from a world of political upheaval, and the omnipresence of artificial intelligence in modern life. The story of acoustic ecology can actually serve as an invaluable model for addressing anxieties produced by the pluralistic, industrialized city: sound is the symptom, ecology is the remedy. In this vein, Schafer famously likens the sounding universe to an orchestration, "in which we are simultaneously the audience, the composers and the performers" (1993, p. 105). While many have critiqued this view as too aesthetically driven and oversimplifying of both problem and solution, Schafer's notion of the soundscape encompasses many more lesser known relational statements that cast acoustic ecology as a complex and flexible system of understanding the socio-political dimensions of sound:

> The electric revolution has this given us new tonal centres of prime unity against which all other sounds are now balanced. [...] to relate all sounds to one that is continuously sounding is a special way of listening. (Schafer, 1977, p. 99)

Embedded in Schafer's thinking are fundamental ideas of relationality, of listening as a form of social relation. In the contemporary moment, sound ethnography is working precisely at those points of tension, and what contributor Vincent Andrisani calls "in-betweenness"—referring to the permeability of sound in the cultural environment of Havana, Cuba. Listening, in fact, can be thought of as a form of civic action that comprises pragmatic, as well as historical forms of listening (Andrisani). In 2010, at the World Forum for Acoustic Ecology in Koli, Finland, contributor Andra McCartney introduced the concept of *ecotonality*. It is derived from the term ecotone, which in the ecological context means a transition area between two ecosystems or biomes, such as between forest and grassland, or the intertidal zones between ocean and shoreline. With this term she expands our thinking beyond Schafer's polarizing concepts of hi-fi and lo-fi soundscapes and their equally polarizing critiques, into the much more complex areas of the transitional and the in-between.

One of the shifts we see in this newer work in acoustic ecology is a central focus on the political, and on redressing social justice. Sometimes I wonder, as we see ideas cycle through time, whether everything in the 1970s seemed so political that to construct listening as political seemed obvious. Of course, in McLuhan's words the rear-view mirror is 20–20 and folks then didn't think of their own positions as politically hierarchical or their own presence as someone else's absence. Today, we have no such privilege.

Randolph:

While Schafer's thinking on sound certainly emphasized relationality, it was also imbued with ways of thinking about these relations informed by a notable bias towards a very particular moment in the history of settler colonialism: that moment when Europeans decided they were going to stay and become Canadians. As Mitchell Akiyama unpacks in his contribution to this volume, much of the work of early acoustic ecology under Schafer romanticizes the state of civilization that made settler colonialism possible along with its attendant technologies: the steam trains, foghorns, and church bells that transported, guided, and gathered people together in this new place (at the expense of so much, and of so many that were already here). In the idea of social permaculture, there is a strain of this drawing of lines in the sands of history, a return to a moment in the past made possible by the privilege of living it from the present, a position that does not demand the truth and reconciliation with indigenous cultures that real harmony with the environment requires. Today's acoustic ecology must function from a different vantage point: one that tells a different story about how the world is and how we want it to be, to acknowledge the privilege upon which the field was founded and find new paths out into society at large.

In the film *This Changes Everything* (2015), Naomi Klein recounts a shift in humanity's thinking about how we relate to this thing we call "the environment," precipitated by the formation of the Royal Society and their anthropocentric notion of humans as striving for mastery over the Earth and its resources, a marked break from indigenous emphasis on sustainability. Klein's premise is that these are simply stories that we tell ourselves about our relationship to everything around us, and that just as indigenous cultures have maintained their ideological positions through the act of storytelling from one generation to the next, so too is our current capitalistic moment the product of the stories that settler cultures have been perpetuating for the past several hundred years. It seems hopeless, at this juncture, to think that we can back away from the course that global warming is charting. For Klein, however, it's a matter of changing the stories we tell ourselves. History shows that stories can and do change regularly. Stories can change everything.

The field of acoustic ecology, as it developed under the direction of R. Murray Schafer in the late 1960s, began by telling a new story about the problems of noise in the modern world. Instead of framing solutions around noise abatement, as was long the norm, particularly at the levels of law and government, Schafer aimed to reframe the issue as a positive: instead of pushing bad noises back, why not create spaces in which good noises proliferate? This was the heart of acoustic design, the idea that if we paid more attention to sound at the planning stages across all of society's projects, we would end up with built environments that functioned sonically on a human scale: the creation of more *livable* environments by attending to sound at the level of the blueprint. To work as Schafer wanted, acoustic design would have to be implemented on macro scales, at the level of urban planning rather than individual buildings, at the level of government rather than private enterprise, at the level of all individuals whose ears need to be cleaned so that we can hear the world as it is and aspire to something better. This hasn't happened for our sound environments, and it isn't happening for the climate either. Schafer's story, positioning audiences as performers and composers, relies too heavily on the need to hear the whole and master it, as a recording engineer might master a recording for optimal playback. As Karin Bijsterveld demonstrates in her analysis of urban music festivals in Europe, the noise abatement approach to problems of sonic encroachment is alive and well in the twenty-first century and, against the grain, has been largely inspired by first wave acoustic ecologists. What if we told a different story about how we relate to our sound environments, and what our sound environments tell us about our relationship to the world? We might begin this story with an alternative perspective on the history of acoustic ecology itself, told by the only woman on the original team of the World Soundscape Project, Hildegard Westerkamp.

Milena:

Hildegard Westerkamp's entry in our volume stands as a unique contribution. Long-time public intellectual and co-founder of the WSP as well as the acoustic ecology movement, Hildegard Westerkamp's piece itself is a dialogue over time, unravelling generational dynamics and the clash of cultural traditions. In one sense, it is a dialogue with the wider academic community of soundscape studies; but it has also been an ongoing private conversation between myself as a friend and student of acoustic ecology, and her as a public voice and mentor of many, a renowned composer, and also a student of life. Westerkamp's piece unearths some unorthodox perspectives bringing the canon of acoustic ecology—deservingly—up to date with the vulnerability and reflexivity of a "beginner's mind." There is no universal humanistic soundscape, or even "balance"—sound reverberates through generations, and individual experiences; through cultures of research and cultures of art, if one is listening. Westerkamp stands out in the landscape of soundscape studies today as someone who tirelessly questions, listens, learns from, and practises radical ethics for emergent themes: themes like social justice, the Anthropocene, intersectionality, and postcolonialism. It is only fitting that the title of her work is "The Disruptive Nature of Listening."

Act 2. Sounding Out (of) the Anthropocene

Randolph:

As I write these words, a monumental joint exhibition is running at the Art Gallery of Ontario and the National Gallery of Canada, titled simply *Anthropocene*, featuring work by Edward Burtynsky, Jennifer Baichwal, and Nicholas de Pencier. The exhibition lays bare not only the state of the world's environmental crises, but also the ways in which world-renowned imagemakers envision these crises, and have been critiqued on several fronts. Burtynsky has come under fire for his tendencies towards distanciation and aestheticization in his famously aerial and abstracting views of large-scale environmental damage, implying that the solution might also come from such a distanced position rather than on the ground (see Fitzpatrick, 2018). Then there is the un-ironically anthropocentric premise of the very notion of the "Anthropocene" itself, a term under much scrutiny across the sciences and the humanities alike. As Karla McManus points out in her essay in the exhibition catalogue (McManus, 2018), debates around the designation "Anthropocene" run along two axes: where to draw the historical line that signals the shift to the "Human Epoch" (the arguments differ by centuries), and whether or not it is correct to designate our current moment as part of the Anthropocene in the first place. Might the world's environmental struggles, so many undoubtedly caused by humans, be bound up in the problem of the position that we humans imagine ourselves to hold here on Earth? There is more than a little hubris behind naming an entire epoch after ourselves, even if the intention behind it is to highlight our negative contributions to the planet. Critics of the designation "Anthropocene" argue that there may yet be things we don't know about our planet, that there are forces at work that might just be carrying on regardless of all our toil and trouble, and that the idea that humans can "fix" the planet is as improbable as the idea that we are so important as to define the age we live in. Further, the sweeping prefix "anthropo" suggests equal implication and responsibility across the human race, which does not reflect the realities of who is most affected by these environmental crises, and who benefits the most for contributing to their causes. We tell ourselves the story of our own prominence as a species in this epoch, but it's not the only story that can be told. This point is essential to any field of study that imagines itself in some way "ecological," for the truly ecological thinker must think outside the bounds of the anthropocentric, whether it be to rethink our current moment or to imagine a way past it.

It will come as no surprise to those of us who work with sound that the *Anthropocene* exhibition is a predominantly visual one, begging the question: what can attention to sound bring to the discourse and its real-world implications? The work of acoustic ecology began as an investigation into the ways in which humanity's lack of attention to sound has created living environments inhospitable to these very same humans (to say nothing of the non-human world). Yet the field has been anthropocentric in its aims, which we can find on two levels: its approach to historicizing sonic environments, and its use of the arts to explore solutions. We might critique the original imaginary of acoustic ecology along the same lines as the designation "Anthropocene" itself: how does one decide where the line should be drawn around the definitive moment upon which a model for understanding the world might be based? And who does this model serve?

Several entries in our volume tackle these issues by way of alternative approaches to the history of acoustic ecology itself and the kinds of work it has produced. Barry Truax, a founding member of the World Soundscape Project, charts several distinct eras in the development of acoustic ecology from the early days with Schafer through to the establishment of the World Forum for Acoustic Ecology in 1993 and the burgeoning of sound studies as a major research area across disciplines in the decades that followed.

What we call acoustic ecology depends heavily on which period we are referring to, and articulating the nuances in these permutations in the present climate is one of the main goals of our volume. As mentioned above, Hildegard Westerkamp offers an alternative pathway through the field, from the perspective of the only female member of the World Soundscape Project and her very different approach to thinking about the core issues of acoustic ecology up to now. Jonathan Sterne tackles the emergence of digital humanities as a field that, he argues, was well developed in the pre-digital media practices of the World Soundscape Project of the 1970s and 1980s. Here Sterne demonstrates how listening to the work of early acoustic ecology can reveal new ways of thinking about broader developments in the humanities, and by extension, the history of technological mediation.

And what of the arts in all this? Can working with sound offer alternatives to the kinds of visual arts privileged by major institutions like the National Gallery of Canada? As part of Truax's history of the field, he offers a guide to various approaches to soundscape composition and its stylistic and technological development over the decades, with an ear towards changing modes to engage with ecological issues. Sterne also helps answer this question by offering an in-depth analysis of two key pieces of soundscape composition: Truax's own *Riverrun* (1986) and Westerkamp's *Kits Beach Soundwalk* (1989). At the same time, Mitchell Akiyama holds up the activist sound art of Ultra-red as a mode of ecological community engagement that offers an alternative to the problems of exclusion that he charts across the early sound works of the World Soundscape Project. Together, these analyses challenge the historicity of acoustic ecology and how its art produces tensions between eco-centric and anthropocentric modes of ecological thinking, offering new ways of understanding how our Anthropocene moment has been lived through sound.

Milena:

I still remember the excitement I felt reading Anja Kanngieser's paper "Geopolitics and the Anthropocene: Five Propositions for Sound." Rather than being centred on Schafer's call for soundscape renovations, the piece starts with an Ursula La Guin quote: a quiet poetic science fiction reminder of micro attention, of listening to the small voices of things. Kanngieser (2015) goes on to propose that the affordances of sound (and listening specifically), "can open space from which to challenge hegemonic and violent forms of subjectivication; forms that have produced this Anthropocene moment" (2015, p. 1). A tall order! And one that speaks most brightly to the epistemological and axiological changes in soundscape thinking from Schafer to the present moment. The imagined significance of sound in five areas of geopolitical thought about the Anthropocene certainly echoes Schafer's aspirations for a better sonic world, a more balanced, harmonious environment that reflects environmental and political problems addressed: less noise, less congestion and masking, better communication across the human world, conservation of natural habitats. Yet still Schafer's project reflects a deep humanistic grounding that has been troubled in the "Anthropocene turn" across the disciplines—that is, acoustic ecology calls on humans to solve problems created by human interference on the planet. Better, more aesthetically beautiful, soundscapes are imagined by Schafer as the ultimate outcome of soundscape work via re-awakening the senses and squaring the balance between recorded, amplified sound on one end, and incidental quotidian soundscapes on the other. These are the acoustic environments that are a by-product of city life: of policy, economics, density, and infrastructure. Critics of the anthropocentric orientation and of the very idea of the "Anthropocene" tend to approach sound, instead, as a metaphor that can constructively inform fraught relationalities between humans and nature, knowledge translations, and political unrest (Kanngieser, 2015): problematics that threaten the balance of life on earth at the crucible of profound human interference. These critics, in other words, are not aiming to reclaim a lost human condition, as was Schafer's drive in early acoustic ecology, but to restore environmental and social justice by de-prioritizing human significance. In other words, they aim to move us out of the Anthropocene, away from top-down colonial, patriarchal, supremacist thinking, and towards a communion of relations:

> Sound is not just about hearing and responding, or communicating. It is about becoming aware of registers that are unfamiliar, inaccessible, and maybe even monstrous; registers that are wholly indifferent to the play of human drama. (Kanngieser, 2015)

With this renewed sensibility in mind, contributors Leah Barclay and Linda O Keeffe adopt an exploratory, artistic sense to the biopolitics of natural soundscapes. In her chapter in this volume, O Keeffe investigates what renewable energy sounds like and how it interacts with existing habitats of soundmaking species. Not coincidentally, O Keeffe is also the co-founder of *WISWOS (Women in Sound/Women on Sound)*: a collective and movement to recognize the marginalized voices of female sound designers, composers and media practitioners. In other words, a critical interest in what Kanngieser calls "imperceptible" soundscapes permeates both O Keeffe's artistic research and her service to the wider sound art community. Similarly, co-founder of *Biosphere Soundscapes* Leah Barclay shares some of her journey recording pristine natural habitats in Australia's rainforest and talks about the way acoustic ecology has informed bringing wildlife recording to urban dwellers. Using contemporary mobile technology, geo-located sound and even hydrophones capturing the underwater rhythms of marine life, Barclay's work is a manifestation of sound's imperceptibility bringing attention to global biodiversity and conservation.

Act 3. Pedagogies of Sound and the Limits of Listening

Milena:

Part of my job is to teach the contemporary versions of all the original acoustic ecology courses in the School of Communication at SFU. Every time I teach one, I take the time to reflect on what I learn from my students. I think they are the toughest critics, the most relevant audience to the ideas of acoustic ecology, and I struggle every time with whether I'm preaching or educating them about it and what value it adds to their lives in a way that makes sense, that moves them in the right direction. One way in which I update ecological thinking in sound has been to permeate it with ideas of the Anthropocene; of climate change; of knowledge production and epistemology; of politics and injustice; of race and gender; of indigeneity. All of these ideas, however, are distant echoes for North American 20-year olds who spend most of their time ensconced in the multisensory electronic boudoir of their mobile devices. So I ask my students these questions: do you think listening is important? Do you think sound is important? Does it matter to you—and if not, why? What *does* actually matter to you? From much probing over the years, I feel that it is the seeming insurmountability of the world's problems—environmental, political and otherwise—that stun young humans and breed apathy, at least among those who fit into the mainstream paradigm enough to afford to be apathetic. Even the city we live in—something I usually make a focal point in my courses, and the original case study of the World Soundscape Project—seems like too complicated a mechanism to try and untangle.

Sound ecology in the city seems then like a futile effort: a theoretical idea that meets dead ends at every turn if we are to truly consider not only pragmatic issues like masking, noise control, and silence, but also social and political issues such as whose noise is policed; who is entitled to silence; which soundmarks are we preserving; how might soundscapes and listening itself be gendered? I myself struggle with these questions and understandably students struggle too. Borrowing from media studies, a focus on representation allows us to think of soundscapes as uniquely representing the internal and invisible struggles and politics of the world. Listening would then bring out these tensions. But where are the limits of listening? As much as cultural sound studies ethnographers would like to listen politically, historically, culturally—and there is a proliferation of recent work across all those aspects—the boundaries between listening as an experiential act and listening as metaphor are at best blurry, and often unspoken.

For instance, it is unclear how marginalized listening positions could be incorporated into city planning, given that cities struggle as it is to accomplish anything more than noise control. How far do we stretch the idea that each person hears a different soundscape dependent on their physiology, ethnology, and cultural position? How precisely do we understand and affect the Anthropocene moment by listening to relations, to the unfamiliar, to the unheard voices of things (Kanngieser, 2015)? Viral videos of the "secret sounds of sunflowers" or the "symphony of ice" seem to re-inscribe the promise that with more sophisticated tools and technologies we may probe and conquer the secrets of the universe: a decidedly anthropocentric aspiration.

The lesson I return to time and time again is the importance of temporality and reflexivity in listening: that it is only through intentional listening to unfolding acoustic environments that sound can have transformative effects. Year after year, my students rediscover and replicate similar tensions in urban soundscapes: themes of gentrification, commercialization, control, and productivity. Yet each time, I see the transformative effects of this type of listening inquiry as they discover something new to them through sound. Connections between black metal and white supremacy; voice and gender on YouTube; mall sound design; and electric car soundscapes are all topics that have emerged from the problematic of acoustic ecology in the classroom: thinking about sounds relationally and noticing historic shifts in sound's aesthetic politics in particular cultural contexts. And in these works, I see the spirit of Hildegard Westerkamp's notion of the disruptive nature of listening and its generative, creative potentials. Paradoxically, "analogue" soundwalking is as popular as ever with initiatives all over the world (Aletta & Xiao, 2018; Gutierez & Grossman, 2017; Radicchi, 2017), while at the same time listening in the city today is profoundly a technological experience. This sentiment is echoed in Taylor and Fernström's practice of "acouscenic listening" in this volume; the activities of their creative brainchild *The Softday Collective* both hail Steph Ceraso's (2014) idea of listening as multi-modal practice, and update participatory forms of soundscape composition through mimicry and (mediated) improvisation. From voice memos to decibel readings, audio mixing and production, mobile media culture has a critical part to play in contemporary acoustic ecology practices and I have been increasingly incorporating technology in the classroom. Students go out listening with pen and paper, as well as their digital devices, and it is my hope to train not only critical media producers, but also citizen-advocates of the "neglected mashup" (Wyse, 2014) that is our shared soundscape.

Did someone say technology? Enter the smart city. The smart city, as we'll see in Sarah Barns' contribution to this volume, is the newest craze, an iteration of the city informed by data, both virtual data that is collected and analysed continuously about human density, habits, consumption and transportation, but also environmental data that should help the city mould, move, and flex towards greater "livability." As part of this trend to "smart"-en environments and infrastructures digital mobile soundmapping has had a strong presence in the last two decades, with more global initiatives such as radioAporee,[1] as well as countless local maps,[2] and an urban planning citizen science tool such as the HushCity[3] app (2017) on the other end. Elsewhere, I've discussed the logics and grammar of soundmapping as potential forms of radical cartography (Droumeva, 2017) that nevertheless require us to move past the illusory democratic promise of digital crowdsourcing. Rather, I have argued, radical soundmaps need to function at the local scale in line with the context and concerns that they are trying to address. It is indeed exciting to see initiatives such as Cities and Memory,[4] or the London Sound Survey[5] out there that have moved so far since the days of the World Soundscape Project (if we consider that a type of soundmap) and rather than trying to encompass "the world," are instead themed, local, and representationally non-Cartesian. What would smart cities sound like then? My answer would be a local orchestration of different stakeholders owning the tools to represent, analyse and communicate about their own sonic experiences in the city.

Randolph:

The SafetiPin app was launched in 2013 to make cities safer for women using digital mapping tools to identify routes that are statistically less prone to street harassment and violence, largely correlated with better lighting (Fleming, 2018). In turn, the app has helped urban planners identify systematically dark areas and improve lighting conditions there, with a concrete effect on the daily experience of their users. This is one example where the relationship between totalizing mastery and individualized experience on the ground has been made clear. Those working in and around sound ecology have made similar maps to chart noise levels as well as historical shifts and community spaces, but it's hard to imagine what the sonic equivalent of the SafetiPin app might be, perhaps one of the reasons why systemic acoustic design has yet to materialize. What kinds of auditory information would have to be mapped to make a city a safer place for people oppressed by violence? The ShotSpotter technology has attempted this for gun violence, identifying when gunshots are fired and using variants on echolocation principles to triangulate the source of the shots and dispatch law enforcement more quickly. Yet as Lilian Radovac argues, the benefits of this technology are dubious and come at the expense of heightened surveillance of targeted communities (CBC Radio, 2018).

Given the ever-increasing propensity for the audiovisual documentation of urban spaces, one of the most pedagogically useful aspects of acoustic ecology in my own media studies classrooms has been the field's interest in tracking and evaluating sounds of the city by way of recording media. While the goals of acoustic design have scarcely materialized according to Schafer's grand ideals, the gradual shift in the work of the WSP from sound recording as objective documentation to a tool for exploring the expressive potential of space reveals also a shift in ways of thinking about lived space and, in particular, the way that communities write their own spaces through living them. The politics of representation are front and centre in today's culture wars over the rights of women, people of colour, and LGBTQ+ communities. It is essential that students learn to understand what is at stake when things look and sound a certain way in the media, and what they can do on a daily basis in their own media production to change things for the better. We are still a visually oriented society, so when students' ears are opened to the sonic dimension of these media, new potentials open up for creating new spaces of inclusion, whether it be an indigenous film production company like Igloolik Isuma Productions in Nunavut, or a social media group tailored to specific communities. These potentials come from experience on the ground, not the totalizing rhetoric of Burtynsky's aerial photography or Schafer's acoustic design.

The realities of street-level oppression for marginalized people force questioning of that privileged staple of urban studies, the flâneur, a figure of the 1800s that Walter Benjamin described as one for whom "the city splits … into its dialectical poles. It opens up to him as a landscape, even as it closed around him as a room" (1999, p. 417). For the classic flâneur, the streets are experienced as interior spaces that can be *inhabited* in a way that emphasizes the lived nature of the social and cultural landscape rather than the alienating detachment so often attributed to urban living (p. 421). While Benjamin's definition holds something of ecology's requirement for experiencing our ecosystems from within, there is also the sense of mastery of the environment that allows this decidedly male figure to participate in culture without the danger that comes with being out in the mix.

Critics like Anne Friedberg (1993), Anke Gleber (1997), and Tom Gunning (1997) have argued for different ways of thinking about flânerie and its offshoots when accounting for female subjectivities and the changing nature of modernity since the nineteenth-century origins of the term. As the very existence of the SafetiPin app attests to, women are not afforded the safety that was required for the flânerie of city streets, so Friedberg points to the emergence of the *flâneuse* in the cinemas and shopping malls of the early 1900s, allowing for a form of mobile spectatorship through the experience of screens and shop windows that operate on a virtual register. This rejigging of urban experience around female subjectivity tied to screen technologies remains couched in visual terms here, raising a parallel critique of the lack of attention to sound in theorizing urban space even across the feminist turn.

Several entries in our collection seek to redress these issues, offering alternative pathways through the theorization of urban experience at the intersection of technologies of mobility. Andra McCartney reassesses the role of traffic sound in defining our experience of the city, and how we listen with more than just our ears, with our whole bodies, and their technological extensions. Here she takes a staple whipping post of acoustic ecology, traffic noise, and rethinks it as a function of the mobility that the urban experience offers. In a similar vein, Heikki Uimonen charts how the radio call-in programme format employed by the Finnish Broadcasting Company has engaged listeners to discuss their auditory preferences on the air, whether lost to the past or present in the moment, establishing a collective virtual soundscape that reflects back on the culture that produces it. The kinds of sounds people latch onto in order to connect with their cities reveal alternatives to the model of signal versus noise established by early acoustic ecology. And Özgün Eylül İşcen discusses how the city of Vancouver, playground of the WSP, has been experienced differently through the mediated ears of migrant peoples. Through a rich analysis of the Vancouver-based video work of Lebanese immigrant Mona Hatoum, İşcen proposes that we adopt Iain Foreman's idea of "acoustic ontology" as a route towards understanding experiences of migration and exile that are not well accounted for in the more holistically oriented rhetoric inherent in the word "ecology." Perhaps, ultimately, this act of renaming is at the core of this entire volume. By acknowledging the limits of listening alone, we come to the limits of the metaphor of ecology itself when used outside of its own ecosystem of the natural sciences. Today's research and practice at the intersections of sound, media, and ecology must find pathways across the rhetoric of inclusion, and account for the mobility required to do so.

Notes

1. See Radio Aporee's daily updated map: https://aporee.org/maps/
2. See, for instance, the Chinatown Soundmap, Montreal Sound Map, Soundseeker, the British Library sound map, and many others.
3. See Open Source Soundscapes: http://www.opensourcesoundscapes.org/hush-city/
4. Created by Stuart Fawkes: https://citiesandmemory.com/
5. Created by Ian Rowes: https://www.soundsurvey.org.uk/

References

Aletta, F., & Xiao, J. (2018). *Handbook of Research on Perception-Driven Approaches to Urban Assessment and Design*. Hershey, PA: IGI Global Press.

Benjamin, W. (1999). The Flâneur. In *The Arcades Project* (H. Eiland & K. McLaughlin, Trans.) (pp. 416–455). Cambridge, MA: The Belknap Press of Harvard University Press.

CBC Radio. (2018, July 30). Tracking the Sound of Gunshots: Is ShotSpotter the Solution to Toronto Gun Violence? *CBC Radio*. Retrieved from https://www.cbc.ca/radio/day6/episode-400-toronto-gun-violence-arctic-wildfires-house-of-frightenstein-johnny-clegg-and-more-1.4761861/tracking-the-sound-of-gunshots-is-shotspotter-the-solution-to-toronto-gun-violence-1.4761881

Ceraso, S. (2014). (Re) Educating the Senses: Multimodal Listening, Bodily Learning, and the Composition of Sonic Experiences. *College English, 77*(2), 102.

Droumeva, M. (2017). Soundmapping as Critical Cartography: Engaging Publics in Listening to the Environment. *Communication and the Public, 2*(4), 335–351.

Fitzpatrick, B. (2018). From Above: Critical Distance, Aerial Views, and Counter-Images. In G. Cammaer, B. Fitzpatrick, & B. Lessard (Eds.), *Critical Distance in Documentary Media*. Cham, Switzerland: Palgrave Macmillan.

Fleming, A. (2018, December). What Would a City That Is Safe for Women Look Like? *The Guardian*. Retrieved from https://www.theguardian.com/cities/2018/dec/13/what-would-a-city-that-is-safe-for-women-look-like?CMP=Share_iOSApp_Other&fbclid=IwAR2Kd-HyTrqnWsIeH3yaItnD_iatFUwG8T5SmcOUrzu1f6rtPADb5hA-qz4

Friedberg, A. (1993). *Window Shopping: Cinema and the Postmodern*. Berkeley, CA: University of California Press.

Gleber, A. (1997). Female Flanerie and the Symphony of the City. In K. von Ankum (Ed.), *Women in the Metropolis: Gender and Modernity in Weimar Culture* (pp. 67–88). Berkeley, CA: University of California Press.

Gunning, T. (1997). From the Kaleidoscope to the X-Ray: Urban spectatorship, Poe, Benjamin, and traffic in Souls (1913). *Wide Angle, 19*(4), 25–61.

Gutierez, A., & Grossman, J. (2017). Cultural Enclaves and Acoustic Territories – A Sonic Study of Urban Development in NYC and Chicago. *Proceedings of Invisible Places: The International Conference on Sound, Urbanism and the Sense of Place*, 7–9 April 2017, São Miguel Island, Azores, Portugal.

Hooks, B. (2000). *Feminism Is for Everybody: Passionate Politics*. Cambridge, MA: South End Press.

Kanngieser, A. M. (2015). Geopolitics and the Anthropocene: Five Propositions for Sound. *GeoHumanities, 1*(1), 80–85.

McManus, K. (2018). How Anthropo-scenic! Concerns and Debates About the Age of the Anthropocene. *Anthropocene: Burtynsky, Baichwal, De Pencier.* Art Gallery of Ontario; Ottawa, ON: National Gallery of Canada; and Bologna, IT: Fonazione MAST, 45–56.

Radicchi, A. (2017). Hush City, A New Mobile Application to Crowdsource and Assess "Everyday Quiet Areas" in Cities. *Proceedings of Invisible Places: the International Conference on Sound, Urbanism and the Sense of Place*, 7–9 April 2017, São Miguel Island, Azores, Portugal.

Schafer, R. M. (1977). *The Tuning of the World*. New York, NY: Knopf. Reprinted as *The Soundscape: Our Sonic Environment and the Tuning of the World*. Rochester, VT: Destiny Books, 1994.

Schafer, R. M. (1993). *Voices of Tyranny, Temples of Silence*. Toronto, Ontario: Arcana Editions.

Westerkamp, H. (2001). Speaking from Inside the Soundscape. In D. Rothenberg & M. Ulvaeus (Eds.), *The Book of Music and Nature* (pp. 143–152). Middletown, CT: Wesleyan University Press.

World Soundscape Project. (1977). *European Sound Diary (Reprinted)*, R. M. Schafer (Ed.). Vancouver, Canada: ARC Publications.

Wyse, P. (2014, March 13). What will our cities sound like in the future? *The Guardian*. Retrieved January 20, 2018, from http://www.theguardian.com/cities/2014/mar/13/sounds-city-technology-urban-centrespeaceful

SECTION I

Acoustic Ecology: Foundations and Critical Responses

CHAPTER 2

Acoustic Ecology and the World Soundscape Project

Barry Truax

Introduction

The emergence of the World Soundscape Project (WSP) at Simon Fraser University (SFU) in the late 1960s and early 1970s has been widely recognized as creating a foundation for the study of "acoustic ecology"[1] with composer and humanist R. Murray Schafer as its "father" (see the cover of the 2012–13 *Soundscape* journal). The key transition in Schafer's thinking that resulted in his call for the creation of the WSP was his change of focus from being anti-noise pollution to a more positive listener-centered approach embodied by the term "soundscape." Schafer's early publications such as *The New Soundscape* (1969) and *The Book of Noise* (1970) were mainly driven by a concern for the rising noise levels in urban areas in the 1960s caused by jet aircraft, construction equipment, motorcycles and amplified music. With his extended essay, *The Music of the Environment* in 1973, later expanded into his book *The Tuning of the World* (Schafer, 1977), he announced that "We must seek a way to make environmental acoustics a *positive* study program … Only a total appreciation of the acoustic environment can give us the resources for improving the orchestration of the world" (Schafer, 1973, p. 3, original emphasis).

B. Truax (✉)
Simon Fraser University, Burnaby, BC, Canada
e-mail: truax@sfu.ca

M. Droumeva, R. Jordan (eds.), *Sound, Media, Ecology*,
Palgrave Studies in Audio-Visual Culture,
https://doi.org/10.1007/978-3-030-16569-7_2

Throughout this period of the 1970s, and even much later, Schafer used musical metaphors to explain his idea of the "world soundscape as a macrocosmic musical composition" with everyone as its composer. His reference to a "tuning" of the soundscape reflected his activist approach, such that critics justifiably characterize his book as "a prescriptive text that is often referred to as a descriptive one" (Kelman, 2010, p. 214). Similarly, his rhetorical use of the audio terms "hi-fi" and "lo-fi" soundscapes was a metaphorical strategy to communicate these new ideas to the public, which for better or worse remains effective to this day, even if they do not stand up to academic scrutiny (McCartney, 2016). Simplified stereotypes attributed to Schafer (such as "nature is good, technology is bad") have often been heard, particularly within social science and cultural studies perspectives that tend to question any value-based distinctions. However, it is hard to underestimate the influence that his ideas have had over the last half century.

By the early 1970s, Schafer had placed the WSP within a new social science school at SFU, then known as the Department of Communication Studies. It quickly became apparent to Schafer's younger colleagues that emerging information-based paradigms and context-centered approaches within the field of communication had a lot to offer soundscape studies. In 1984, Barry Truax consolidated such an approach with his book *Acoustic Communication*. Limiting his model only to the human perspective, Truax described the soundscape as a "complex system … that sound creates between people and their environment" (Truax, 1984, p. xi). That is, he and the WSP group of researchers emphasized the perceptual, cognitive, qualitative and social aspects of the soundscape rather than the standard quantitative approach to the acoustic environment that was and is still dominant in the noise abatement industry. At the macro level, Schafer's soundscape approach preferred to ally itself with emerging ecological movements that were particularly strong on Canada's West Coast (the original home of Greenpeace, for instance), by suggesting that soundscape ecology be added to the environmentalist agenda. The theoretical underpinnings of such an ecological approach were that of systems theory where the acoustic environment could be regarded as a complex system of interacting elements, rather than isolated acoustic phenomena. This concept of the soundscape as an all-encompassing understanding of the sonic environment complemented and extended the ecological model, as it does to this day.

Theory and Practice: The Research Work of the WSP

Admittedly, Schafer and his team had no specific training in environmental sound analysis and evaluation, and in fact at that time, there were few if any qualitative approaches to such work, and certainly none that were listener centered. Techniques had to be developed in the field in response to what was found there. The group relied on their aural abilities, mostly trained through both acoustic and electroacoustic music production, and went out with a high-quality stereo Nagra recorder, the film industry standard then, a sound level meter and an endless inquisitiveness to document the soundscape and people's reactions to and awareness of it. Today their entire archive of raw recordings and finished publications is available online as the WSP Database. In particular, their recordings from the 1970s, first in Vancouver (1972–73), and then across Canada in 1973 and several countries in Europe in 1975, resulted in a collection of nearly 200 hours of stereo recordings, all well documented and described for their perceptual qualities. Besides the chronological catalog, the recordings are categorized by a three-level subject index, where the top five headings are Sounds of Nature, Human Sounds, Sounds and Society, Mechanical Sounds and Sounds as Indicators, with the third and fourth of these categories being the largest by far. Two additional collections of Vancouver recordings from the 1990s and 2010–11 round out the current holdings, thus giving the city a 40-year longitudinal documentation, aided by interactive maps and links between similar recording sites over that period.

From the perspective of acoustic ecology, the most important research of the WSP was their initial study of *The Vancouver Soundscape*, issued as a booklet and two LP's with that title in 1973, followed by the *Five Village Soundscapes* study that was carried out in 1975 in five European villages and published in 1977 as a booklet with sound examples on cassette tape. A group of Finnish researchers revisited those villages around 2000, and added a Finnish one, with their results published as *Acoustic Environments in Change* in 2009 (Järviluoma et al., 2009) that includes a reprint of the 1975 WSP study, along with four CD's of audio recordings from each village and time period. Similarly, the 1973 Vancouver release has been republished, along with recordings and compositions made in the 1990s, as a double CD (*The Vancouver Soundscape 1973/Soundscape Vancouver 1996*) by Cambridge Street Publishing.

Besides the recordings, the WSP team established "soundwalking" as a simple technique to evaluate the perceptual and qualitative aspects of a soundscape through a listening walk. They also documented historical aspects of the communities studied through archival research and "earwitness" accounts, similar to "oral history" but expanded to comprise more of an "aural history" approach with long-term residents by concentrating on sounds heard and remembered. Soundwalks and earwitness accounts were sometimes combined into what is known today as a "memory walk," where the subject and interviewer revisit the locales of remembered experiences (Järviluoma, 2017). Sound levels were usually documented with some translated into isobel maps (showing equal sound level contours), and often supplemented by sound counts, estimates of acoustic profiles (i.e., how far a community sound could be heard) and timelines of how frequently and long a prevalent sound was audible. The community surveys included examples of the perceptually based concepts of sound signals (those conveying specific information), keynote sounds (those in the background of listening attention), and "soundmarks" (sounds of unique historical and cultural importance in a community). The soundmark concept opened their analysis to the inclusion of social and cultural aspects, and over time, it has been generalized to any sound that locals may deem of particular importance. Noise sources were given special attention for their impact on listening habits in the community and other effects, but in general, the WSP was constructing an account of what they called the "acoustic community."

Truax has described the acoustic community as "any soundscape in which acoustic information plays a pervasive role in the lives of the inhabitants (no matter how the commonality of such people is understood)" (Truax, 2001, p. 66). Given that the WSP's original definition of soundscape included electroacoustic sounds (and media) considered as an environment, the acoustic community concept may be extended into the much differently functioning "electroacoustic community," which is less constrained by physical and temporal continuities and often functions as a surrogate audio environment. In the WSP's 1978 *Handbook for Acoustic Ecology* (Truax, 1978, 1999), the WSP collectively defined the soundscape—as distinct from the sonic environment—as placing the "emphasis on the way it is perceived and understood by the individual, or by a society," thereby raising the difficult methodological question of dealing with subjectivity as well as cognitive processing. Interestingly, the current ISO Working Group 54 on soundscape has repeated and expanded that definition in its proposed standard, now internationally approved, with the intent of guiding qualitative research practices in soundscape evaluation that give context to the more traditional quantitative, objective data.

Arguably, the most fundamental aspect of the acoustic community, and acoustic ecology in general, is the concept of an "acoustic space" which sound creates through its interaction over time with the physical environment. In fact, the perception of sound sources cannot be separated from the influence of the physical space in which they are produced because the acoustic characteristics of the source are colored by that space. Truax (2017a) has proposed the term "dual processing" to describe the simultaneous auditory perception of two sonic dimensions where information about each is effortlessly extracted by everyday listening habits, even if they are not analytically focused by foreground listening. Moreover, there can be an interaction between the perceived "inner acoustic space" of a sound, that is, its volume and perceived magnitude, and the physical space into which it is projected, the "outer acoustic space." This everyday perceptual evaluation is that of how large or energetic a source is, and its distance from the listener. One aspect of the balanced functioning of an acoustic ecology is an evaluation of whether the perceptual magnitude of sound sources is appropriate to the acoustic space they inhabit. This aural ability is fundamental to one's orientation (or disorientation) within a space at the individual level, and to the definition of the acoustic community as a whole.

In 1974, an interesting link emerged between the WSP's soundscape studies and the early work of Bernie Krause in the bioacoustic world, in terms of his proposed "acoustic niche hypothesis" (now known as the ANH in soundscape ecology). Spectrograms of Krause's recordings, particularly in tropical and sub-tropical areas, showed that species' soundmaking was generally confined to non-overlapping spectral bands, thereby creating an acoustic habitat, not merely a biological one (Krause, 1987, 2012). In other words, each species used different frequency ranges such that their spectral components were interleaved in the "niches" of the frequency spectrum. In addition, there could be temporal niches as species alternated their soundmaking over time.

Human auditory perception works on an even finer resolution (than those of animals) of different frequency ranges called the "critical bandwidth." That is, groups of hair cells in the inner ear respond to different frequency ranges (in general, a little less than a quarter octave in the middle range). The critical bands are spaced along the basilar membrane inside the cochlea, whose response forms a "spectral envelope" that serves as a spatial pattern of resonance regions, projected onto the auditory cortex. This spatial pattern is distinct from the much more finely tuned temporal analysis of periodicity that results in pitch sensitivity of less than a one percent frequency deviation. The masking property of one sound over another depends not merely on it having more energy, but also that its energy occupies the same critical bandwidth in its spectrum. Therefore, at the most basic level, an acoustic ecology depends on the balance of sounds in both the spectral domain and the temporal. The connection of the ANH to the WSP work, although not actually recognized at the time, came with their 24-hour "Summer Solstice" recordings in June 1974, on the rural grounds of Westminster Abbey near Mission, BC, where the birds and frogs around a pond formed an ecological micro-environment. An edited version of these recordings consisting of about two minutes per original hour was broadcast on the CBC Radio in stereo as part of the *Soundscapes of Canada* 10-part series in October 1974, the other programs being based on the cross-Canada recording tour. Aural and graphic analysis of this soundscape showed both the temporal and spectral balance predicted by the ANH. As impressive as this balanced soundscape was to the recordists, the question remained as to how a human community could function as positively, a question that began to be answered by the *Five Village Soundscapes* study the following year.

The analysis of a bioacoustic ecology will likely be concerned with factual issues of how it functions with respect to survival and mating, food sources and predators, communication between species and individual inhabitants, as well as territorial definition and its maintenance. One can readily identify analogous functions within human acoustic communities, but what is more challenging to understand from an ecological perspective are the affective and subjective aspects of the human response to sounds, as well as the cognitive processes that individuals develop to derive meaning from their soundscapes. Recent qualitative research in this area, particularly in Europe, has made progress in the former area with participant evaluation (often during or after a soundwalk) of the perceived affective quality of a soundscape, using such adjectives as pleasant, chaotic, vibrant, uneventful, calm, annoying, eventful or monotonous, with overall assessments such as satisfaction, appropriateness, and other valanced (positive or negative) criteria.

It is much more challenging to obtain an overview of what sounds mean to listeners in a soundscape, partly because this cognitive level is habitual and not usually put into words. An in-depth interview by a skilled interviewer may tease out these meanings, perhaps in situ with a memory walk, or when recordings of specific soundscapes are played back during an interview. In his doctoral thesis work at SFU, Christopher Smith (1993) identified four aspects of listeners' acoustic sense of place in his in-depth interviews: (1) Sound as context; (2) Sound as information; (3) Sound and feeling; (4) Sound and memory. This project and related ones show the depth and complexity of the human experience of the soundscape, and add insights into how that relationship is constantly, if slowly, changing over time.

The WSP study of the five European villages in 1975, together with the Finnish study 25 years later of these same villages (Järviluoma et al., 2009)—all of which had experienced significant changes in the interim—began to unravel many of these dimensions within the limits of the durations of their visits and language barriers with the inhabitants in four of the villages. However, the *Five Village Soundscapes* (WSP, 1977b) offered three general criteria in its concluding chapter that seemed applicable to all villages. It should be noted that there were significant differences in the character and history of each village, and in fact, each formed its own model of what constitutes an acoustic community. Moreover, technological innovations had come to each village at different points over the previous century. The villages included a modern manufacturing village in southern Sweden (Skruv, where a glassworks, brewery and two other industries were located), a traditional farming village in Germany (Bissingen, where a textile factory was also operating), a fishing village in Brittany (Lesconil), a mountain village in Italy (Cembra, experiencing declining population as people migrated to the city), and an academy town in Scotland (Dollar, on major highway and rail routes, but with strong links to the past). From the sparse Scandinavian village soundscape, to the French and Italian ones centered on human activity, and those experiencing the effects of industrial and technological trends, these villages presented a range of examples where sound still played a largely positive and pervasive role in the definition and experience of the community.

The Finnish team that revisited these villages 25 years later found significant changes in those villages. The fishing fleet in Lesconil had moved to a different harbor, though the Breton traditions of the region were flourishing; Cembra had adopted various strategies to attract tourists to their traditional way of life (ironically portrayed as an escape from the cities which had drawn many locals away); others were evolving more slowly, and less dramatically but along the lines of modernization. As referred to earlier, the WSP study drew several conclusions about the character of the acoustic communities they studied, some of which were identifiable within the language of ecology. The first was the *variety* of sonic "species" that were heard in these villages. Instead of the soundscape being dominated by a few, acoustically powerful sounds, these communities exhibited a wide range of sounds that could be heard on soundwalks throughout the village. The writers noted that this conclusion was rather the opposite of the cliché that life in the modern city is "complex"—culturally perhaps, but not necessarily aurally. The second characteristic of the acoustic community was the *complexity* of information gleaned by the inhabitants from everyday sounds. This was often noted by the researchers, but could only be explained by the inhabitants who had the requisite knowledge of context, culture and history. Although the experience of sound in such a community is shared, the interpretations and reactions of the inhabitants can be quite varied, depending on personal experiences, preferences and relationships to the community's power structures. Finally, a third, more explanatory criterion was identified, namely, how the variety of sounds were maintained in an ecological *balance*. What kept the variety (and magnitude) of sounds from overwhelming each other and producing a chaotic soundscape?

Some basic factors in the perceptual balance of individual sounds have already been mentioned such as non-overlapping spectral qualities, a balance of loudness levels and durations over time, and the interaction and perceived appropriateness of the magnitude of sounds within the given physical spaces. Such relationships have presumably co-evolved with the human auditory system and resulted in higher level auditory perceptions such as what Albert Bregman (1990) terms "auditory scene analysis" where different patterns or "streams" can be simultaneously separated by listeners (although most research in this area is confined to speech and music). The types of acoustic spaces described here can be placed along a continuum:

Anechoic Spaces — Interactive Spaces — Diffuse Sound Field

Briefly, anechoic spaces are those that maximize absorption of acoustic energy and hence provide little sense of space other than distance. Diffuse sound fields, at the other extreme, are those that maximize acoustic reflections and build up a relatively undifferentiated sound field through excess reverberation in enclosed and semi-enclosed spaces. One may say that in anechoic conditions, "the sound is the sound," whereas in diffuse sound fields, "the sound is the space," and in general, both are disorienting, although meditative type practices may allow the listener to adapt. Whereas anechoic conditions generally have to be specifically constructed for research purposes, diffuse sound fields occur frequently in built environments (e.g., a swimming pool or gymnasium with many reflective surfaces, and in downtown corridors where tall buildings trap sound). Both types of extreme acoustic spaces seem to isolate the listener, whereas in the middle of the continuum, listeners are freer to interact with their acoustic spaces, navigate effectively through them and generally feel a sense of agency.

Acoustic communities, when regarded on a macro level according to the WSP model, seem to have evolved (as in the examples described above) according to several balancing factors related to physical space, time (e.g., rhythms and cycles) and social practice. In many cases, economic, social and cultural factors have determined the physical design and layout of a village, town or city, but each decision has an acoustic impact, and so it might be more accurate to say that there is a co-evolution between acoustic and cultural developments. Clearly both aspects need to be examined together as part of what we are calling an ecological system. The acoustical science that is well established today, such as architectural acoustics, sound propagation models and acoustical engineering, has only emerged in the last 150 years, and today, researchers often attempt to uncover the acoustic principles behind specifically effective examples of good acoustic design from past millennia.

The type of co-evolution described here tends to produce a well-functioning soundscape, but it is also one that is vulnerable to change when the balancing forces are disrupted. Noise issues can arise locally or on a larger scale, which along with other environmental stressors can necessitate coping or adaptive strategies by individuals, based on personal factors such as salience, information load, the sense of personal control, as well as prevailing social, economic and political conditions (Campbell, 1983). One of the difficulties of the subjective soundscape model is that, while changes in individual listening habits as advocated by the WSP may trigger the desire for change and a sense of empowerment, it will likely be difficult to reach a broad social consensus, particularly given the human ability to adapt to and ignore even what are negative influences. Therefore, both top-down and bottom-up strategies for acoustic design need to be developed, just as with other environmental risks that society faces today.

The lesson one learns from the ecological perspective is that all elements are deeply intertwined and act as an integrated system closely tied to cultural contexts.

Acoustic Ecology and the Emergence of Sound Studies

Since the pioneering work of the WSP in the 1970s which has laid a foundation for the concepts of soundscape and acoustic ecology, a great deal has changed, most obviously technologically, but also in our understanding and practice of sound research and creation. On the academic front, one of the most notable (and encouraging) developments is the emergence of "sound studies" as an interdiscipline, though generally based in the social sciences. Among the many facets of this widely accepted umbrella term, we now have *The Sound Studies Reader*, edited by Jonathan Sterne (2012a) with 45 foundational articles and *The Oxford Handbook of Sound Studies* (2012) edited by Trevor Pinch and Karin Bijsterveld; there are also several online and print journals, such as *Sound Studies* and the *Journal of Sonic Studies*, a European Sound Studies Association, various international conferences, graduate programs, research projects and so on. As Sterne remarks in his introductory article, "Sound studies' challenge is to think across sounds, to consider sonic phenomena in relationship to one another—*as types of sonic phenomena rather than as things-in-themselves*—whether they be music, voices, listening, media, buildings, performances, or another path into sonic life" (Sterne, 2012a, p. 3, original emphasis).

In the 1970s, there were remarkably few texts that demonstrated an interdisciplinary approach to sound, whose study had become fragmented into numerous academic and applied fields across the fields of science, technology, medicine, the arts, linguistics, the humanities and social sciences, each with their own terminology, theoretical and applied models. The exceptions that proved this rule included Don Ihde's *Listening and Voice* (1976, 2007) that described those topics from a phenomenological perspective (which he has since enlarged to deal with technologies of perception), Schafer's *Tuning of the World* (1977), and Truax's edited *Handbook for Acoustic Ecology* (1978, 1999) that dealt with the disciplinary terminology involved in sound studies. In contrast, numerous books on noise pollution were based on standard acoustics and applied acoustical engineering, with little insight into listening practices and cultural attitudes on the individual and community level, or the political economy of noise regulation. Audio media texts were generally technical or dealt anecdotally with media history, and were outnumbered by more visually based studies. One exception was Tony Schwartz's *The Responsive Chord* (1973) that dealt with media and advertising in terms of sound. No interdisciplinary journals for sound studies existed, and most anthologies were collections of discipline-based studies.

If we jump to the current century, we are greeted by a veritable explosion of interdisciplinary sound studies texts, far too numerous to summarize. With apologies to those who could have been easily included as examples, here is a representative sampling of books, chosen for the importance of their impact and their relationship to the themes discussed here. One remarkable class of documents is those that have dealt significantly with historical soundscapes through careful research into the available documents, even in eras where no sound recordings exist. Emily Thompson's landmark *The Soundscape of Modernity* (2002) grounds her study in the history of science with the career of Wallace Sabine in establishing the modern framework of quantitative architectural acoustics, from the late nineteenth-century acoustic design of Boston Symphony Hall to the 1930s creation of Radio City Music Hall with its audio-reinforced acoustics. She contrasts these developments in the control and design of sound in enclosed spaces to the failure, as she describes it, of anti-noise campaigns in New York City, one of the first cities to study the problem systematically and introduce bylaws. Into this mix of influences, she adds the introduction of reproduced sound via radio and recordings, where advances in sound absorption techniques allowed acoustically isolated sound studios to minimize reverberation that worked against the comprehension of monophonic sound. She also adds examples of the use of "noise" in avant-garde music, to round out the complex interplay of science, technology and culture during this transition to the modern era. Jonathan Sterne also bases his work in critical, historical analysis, such as *The Audible Past* (2003) which traces the acoustic transduction model that informed modern sound reproduction, and the recent *MP3: The Meaning of a Format* (2012b), that unravels the twentieth-century interplay of psychoacoustics, audio transmission and industrial format standards, all of which ground this modern mode of audio consumption.

Equally remarkable are texts that deal with more remote soundscape history, such as Bruce Smith's *The Acoustic World of Early Modern England* (1999), which attempts to reconstruct Elizabethan soundscapes from documentary evidence, with the acoustics of Shakespeare's Globe Theatre as its representative example. Alain Corbin documents the theme of bells in *Village Bells: Sound and Meaning in the Nineteenth-Century French Countryside* (1998). Arguably, the most comprehensive historical study of sound, but in this case noise, is Hillel Schwartz's encyclopedic *Making Noise: From Babel to the Big Bang and Beyond* (2011) where in 859 pages of literary storytelling and engaging case studies, most of which one has never encountered before, Schwartz uses the concept of noise as part of the "history and anthropology of the senses, where noise runs the gamut from the Natural through the Sociable back around to the Unsociable and Ill-Natured" (Schwartz, 2011, p. 23). Karin Bijsterveld, whose work is based in science and technology studies, has documented the intricate problems of noise abatement and its evolution from 1875 to 1975, as intertwined with politics and public discourse in *Mechanical Sound* (2008). More recently, her edited anthology *Soundscapes of the Urban Past* (2013) creates a framework for understanding various media representations of the urban soundscape.

In fact, edited anthologies are a popular form of bringing together related research around a multi-disciplinary topic of study, such as Michael Bull and Les Back's *The Auditory Culture Reader* (2003), *The Routledge Companion to Sounding Art* (2017), and the forthcoming *Oxford Handbook of Sound and Imagination* (2019), among others. All of this work seems to have the greatest relevance when it follows Jonathan Sterne's suggested characteristics of interdisciplinarity, reflexivity, historicity and criticality (Sterne, 2012a, p. 5). He also suggests that its practitioners, coming from numerous backgrounds, should be called "sound students," since there is so much to learn. One of the typical gaps in the training of current scholars and students, particularly in the humanities and social sciences, is a lack of knowledge about acoustics and psychoacoustics. This is a problem in any interdisciplinary undertaking where one draws on material from other areas where one has little training, but it is particularly unfortunate in sound studies where so much knowledge exists about the core subject of the field. Authors often make basic mistakes in terminology and lack factual evidence that in many cases would actually strengthen their argument. Admittedly, the dominant scientific model is that of energy and signal transmission with resultant "effects," that is often woefully ignorant of social and cultural implications, but it is also possible to expand those models, for instance, with the information and context-based acoustic communication model (Truax, 2001) and the closely related acoustic ecology concept.

Another problem is the persistence of simplified concepts that may have been useful in the past, but today are in need of elaboration because of new technology or more advanced scientific research. For instance, received ideas about music in Western culture are clearly inadequate to deal with the current functional uses to which it is put, the political economy of the global media industry with music as a consumer product, not to mention the complex interplay of its cultural, counter-cultural, multicultural and commercial cultural forms. Fortunately, some significant literature has begun to fill this gap such as Tia DeNora's 2000 book, *Music in Everyday Life*, and Georgina Born's 2013 anthology, *Music, Sound and Space*.

A subtler example comes from the listening practices associated with music technology in particular, although these usually interact with other everyday types of listening. Historical accounts often implicitly assume that listening behavior is invariant and not susceptible to its own evolution in tandem with technological and cultural shifts. Admittedly, listening habits are more difficult to document than technology, but it is clear that, for instance, listening habits have co-evolved in ways that are more analytical to the inherent qualities of sound, and at the same time in ways that treat sound as a surrogate, accompaniment environment which stays in the background of one's attention. In other words, the range of listening habits has enlarged, but discourse on listening often relies on the traditional model of listening as only involving foreground attention (i.e., an active-passive dichotomy). This ignores the essential role of cognitive functioning to process incoming auditory input not deemed to be salient in the given situation and keep it in the background of one's attention (ranging from processing during sleep to situations where we are distracted by other tasks). The contemporary understanding of listening needs to recognize that all levels of attention are still listening processes. Cultural studies scholars, such as Carolyn Birdsall (2012) in her otherwise excellent account of Nazi era soundscapes, often claim to find an implicit hierarchy of values (and bias) in the listening continuum from foreground to background levels of attention. One can critically analyze the political-economic implications of media consumption that capitalize on distracted listening, but it doesn't mean that such listening habits are not functional, nor that an acoustic ecological approach wouldn't encourage more acute listening practices to ensure the value that sonic experience gives to our lives and the environment as a whole.

Soundscape Composition

The creative side of environmental sound studies is generally termed "soundscape composition," and although it has obvious connections to contemporary music, it can also include projects that bridge the artistic with the scientific, social and environmental (Truax, 1996, 2002, 2008; Westerkamp, 2002). Most often, the work seems to be about place and utilizes environmental sound (Norman, 2012), on site or via environmental sound data or recordings, but it can also be even broader than that. Truax has proposed "context-based composition" to refer to work where "real-world contexts inform the design and composition… at every level" (Truax, 2017b, p. 1). Placing the emphasis on the relationship to real-world contexts distinguishes this work from historical Western art music practice (including electroacoustic music) which minimizes contextual references, or relegates them to secondary status (e.g., program music), in favor of an abstract treatment of musical sound, that is, sounds related only to each other through various organizational schemes. It also opens up and acknowledges the creative work of those whose primary intent is not artistic, but who may be thought of as sonic designers. However, within the electroacoustic music community, interest is increasing, as witnessed, for instance, by the fact that two calls for theme issues of *Organised Sound* devoted to "Sound, Listening and Place," volumes 16(3) and 17(3), and "Soundscape Composition as Context-based Creation," volumes 22(1) and 23(1), each resulted in a record number of submissions and two full issues.

Soundscape composition largely draws on two traditions that have evolved over the past century, namely, electroacoustic music and radiophonic sound production, even as it differs from and extends these practices (Black, 2014). Here we will comment on those practices that use environmental sound compositionally. One common confusion is the relation between soundscape composition and the tradition of *musique concrète*, which originated in France in the 1940s with Pierre Schaeffer whose early "noise studies" included the *Étude aux chemins de fer* (1948), based on recordings at the Gare des Batignolles. Partly because of the difficulty of field recording then, but more significantly, because Schaeffer was troubled by the "anecdotal" nature of environmental sounds, subsequent work was mainly based on studio recordings of single sound sources, recordings that could be fixed (initially on disk, and later on tape) as a "sound object." By recording these materials in an acoustically isolated environment and close miking the source, traces of environmental ambience were minimized, and the resultant sound treated as an "object of perception," a much different approach from recording environmentally contextualized "sound events" in soundscape work.

Schaeffer's work developed a psychoacoustically based practice of perceptual sound classification known as spectromorphology, literally "sound shapes" describing the evolution of spectra over time. Compositionally, these de-contextualized sounds were musically recontextualized in a style known as "acousmatic music," a term whose derivation indicates the lack of a visual sound source, except the multiple loudspeakers over which it is usually heard. However, Simon Emmerson (1986) points out that it is not the reliance on acoustic, rather than synthesized, materials that distinguishes this genre, but rather its syntactical organization strategies based on inherent sonic qualities and not independent abstract structures. Coincidentally, Luc Ferrari, a French composer of Italian heritage who worked with Schaeffer, challenged the norms of acousmatic music with his *Presque rien* ("almost nothing") works in the 1970s and 1980s, which lightly manipulated environmental recordings and presented them as authored musical compositions. Although acousmatic music and soundscape composition both put the emphasis on listening and perform their works over multi-channel loudspeaker systems to create a multidimensional sound space, there remains a profound difference for listeners in terms of their recognition of environmental sound sources and the contextual knowledge they are encouraged to bring to the listening experience.

In the 1970s, the radiophonic community would not have regarded Ferrari's work as radical, given their longstanding use of environmental sound in radio documentaries and dramas. On the other hand, they would not have presented their work as "sound art," even though it was usually well crafted and composed. In the late 1920s, for instance, the German experimental filmmaker and radio producer, Walther Ruttmann began recording environmental sound on the optical soundtrack of film stock. Unfortunately, this was not possible for his magnificent 1927 silent film *Berlin, die Sinfonie der Großstadt* (Berlin, Symphony of a Great City), which documents daily life in the city for an entire day in five "acts," although the orchestral accompaniment often mirrors its soundscape. Its circadian structure established a compositional norm, along with other "day in the life of a city" films of that period. Ruttmann followed this with *Weekend* (1930), a "sound film" in six sections recorded on the optical film soundtrack. Its structure included the sections (in translation), Jazz of

Work, Quitting Time, Journey to the Country, Pastorale, Return to Work, and a short coda mirroring the initial section. It intricately edited numerous short clips of material into complex rhythmic patterns, creating a coherent montage rivaling the French *concrète* work 20 years later, and relying on source recognizability and a time structure familiar to listeners.

Another tradition based on environmental sound is that of nature recordings. The German/British broadcaster and sound recordist, Ludwig Karl Koch, began recording animal sounds in 1929 (having made the first known recording of birdsong in 1889 with a phonograph). He published various "sound books" (i.e., a book with a disk) in both Germany and England in the 1930s. This practice created a catalog that isolated and identified each species, consistent with scientifically motivated field recording. However, Koch's most striking work is *Hunting By Ear* (1937), an audio book on fox hunting, with two 78 rpm disks (Drever, 2017). The first is a catalog of the various cries of the master of the hounds, but the second is a condensation (via editing) of an actual 40-minute foxhunt into a 6-minute montage he called a "composite sound-picture." Given his cumbersome equipment, this production was a remarkable documentation of a macro sonic event. Another historical precedent is the work of the Canadian recordist, William W. H. "Bill" Gunn, as in his popular 1955 LP, *A Day in Algonquin Park*, constructed as a montage of various recordings, creating what Cameron and Rogalsky (2017) call the "circadian audio portrait."

The American sound recordist, advertising designer and media theorist, Tony Schwartz, is well known for his field recordings from New York City, many issued on the Folkways label in the 1950s and 1960s. Most are presented as a set of edited tracks, featuring the multi-cultural traditions of the city, particularly children's street games, along with some typical environmental sounds. His compilation album, *Sounds of My City*, is narrated and provides a good overview of his work, now largely archived in the Smithsonian Institute. The British born Imbert Orchard recorded nearly a thousand interviews with pioneers in British Columbia, and produced many radio documentaries, such as his *Living Memory* series for the CBC. His practice evolved from being based in recorded voices (oral history) toward a concept of "aural history" that included the wider soundscape context. He also theorized "levels of remove" in the documentary tradition, moving from actuality (original field recordings), to running commentary which he expertly provided in the field, then to recall, and in historical cases, to re-enactment based on written texts. He often interwove those materials into a seamless whole without a master narration, except as provided by his own voice.

Glenn Gould extended the freedom of using studio editing and mixing with his famous *Solitude Trilogy* of works in the 1960s and 1970s, including the monophonic *Idea of North*, the stereophonic *The Latecomers* about Newfoundland, and *The Quiet in the Land* about the Mennonite communities in Manitoba and Ontario, a program that features the most complex layering of religious and secular material to illustrate its major theme. Finally, the extremely strong radio documentary and drama traditions in the UK, Europe and elsewhere should be acknowledged. Although most of this work relies on studio recordings, some remarkable exceptions use environmental sound creatively. Between 1958 and 1964, the BBC commissioned an ambitious series of "Radio Ballads" from producer Charles Parker and ethnographers Ewan MacColl and Peggy Seeger, which combined field recordings and interviews with newly composed folk music based on those interviews, which were then tightly synchronized in the mixing studio with the live performers of the music. The German radio producer Peter Braun extensively used high-quality stereo field recordings in his 1973 Prix d'Italia documentary *Glocken in Europa* (Bells in Europe) to illustrate both the lifecycle of the community and that of the continent's bells, which had regularly been melted down for military cannons and artillery, and later recast as bells. Today, broadcast documentaries and innovative radio work are still practiced in the public non-commercial media, and increasingly find a home in the podcast.

The range of current work operating within the soundscape composition paradigm is rather large and can be positioned along a continuum of practices. At one end there is extensive work in sonification where real-world data is mapped onto or interpreted as sound, sometimes termed "auditory display." Applications range from artistic projects driven by scientific data in installations, sculptures, fixed media works and concert pieces, such as the EcoSono projects by Matthew Burtner (2011, 2017), to those where artistically informed design strategies are used to communicate scientific data to the public, such as issues involving environmental risks (Polli, 2016, 2017). In the middle of the continuum, we find environmental recordings being used with minimal manipulation and processing for a variety of documentary purposes, identified as "phonography" (Drever, 2002, 2017). These practices are often based in environmental bioacoustics or the practice of sonic ethnography, with all of the attendant questions of representation, reflexivity and subject involvement. When greater compositional manipulation is involved, we find the auditory creation of "virtual soundscapes," initially in stereo format and now increasingly reproduced with massive multi-channel playback systems in specialized environments that convincingly reproduce a three-dimensional, immersive soundscape (Truax, 2017a).

Soundscape compositional work continues in the Sonic Research Studio at Simon Fraser University, often using recordings from the WSP library available online, and processed sounds digitally spatialized in eight or more channels with the TiMax2 SoundHub, developed locally and marketed by Outboard Inc. (UK). Hildegard Westerkamp's soundscape-themed works dating to the late 1970s also include eight-channel works, such as *Into The Labyrinth* (2000), from her trilogy of India soundscape compositions, and *Für Dich – For You* (2005), based on Rilke's *Liebes-Lied* heard in both German and English, within an immersive soundscape suggestive of North Germany, Hildegard's birthplace, and Canada's West Coast, her home for many years. Both works evoke poetic imagery associated with their themes through soundscape materials and their transformations.

Barry Truax's eight-channel soundscape compositions begin with *Pendlerdrøm* (1997), which follows a commuter's ride home in Copenhagen by sequencing realistic train station soundscapes with "daydreams" of processed materials previously heard in the realistic sections, thereby simulating a common experience of "tuning out" in a familiar soundscape. *Island* (2000) suggests a visit to an imaginary island where hyper-realistic soundscapes are mixed with processed versions that evoke a level of mystery and perhaps magic. *Temple* (2002) creates a choir in a highly reverberant space where the voices are extended by a resonance similar to reverberation, until at the end only harmonics remain in this spiritual soundscape. Although not evocative of a real space, *The Shaman Ascending* (2004–05) does something similar to *Temple* with chanting voices reminiscent of Inuit throat singing. *Prospero's Voyage* (2004) suggests a scenario where Prospero leaves his magical island, at the end of Shakespeare's *The Tempest*, but does so underwater. *Chalice Well* (2009) creates a completely imaginary soundscape in the mythical caverns beneath the sacred site in Glastonbury, with visits to the Chamber of the Feminine, the Glass Chamber, and an encounter with underworld demons and the Holy Grail. As a counterpart to the water imagery of *Chalice Well*, *Fire Spirits* (2010) deals with fire imagery, and *Aeolian Voices* (2013) with wind. To complete the cycle of the four elements, *Earth and Steel* (2013) recreates an historical steel shipbuilding site and enhances its metallic resonances. *The Garden of Sonic Delights* (2015–16) suggests a visit to a virtual garden with imaginary sonic species, and *Ocean Deep* (2017) creates an imagined descent to the depths of the ocean. These 11 works explore the relation between real-world soundscapes and their imaginary extensions into a virtual world, often heard in multiples of eight channels with a vertical dimension included in venues equipped with speakers at different heights.

The range of work described above, linking electroacoustic music traditions to those in the radiophonic sphere, as broad as that is, can also be linked to current work in game sound, film sound design, and certain aspects of sound art, particularly those that create sonic environments experienced as if they were actual soundscapes. These design contexts include those simulating the psychoacoustics of a 3D soundscape, mainly in the game world, using binaural and other cues to position the listener in a realistic space, through to those relying on traditional studio mixing techniques to combine sound events with an ambient background to create a believable soundscape. In such cases, sound design and soundscape design become integrated and their distinctions blurred, whether the purpose is artistic or commercial. The accompanying listening strategies are a combination of everyday norms, such as the WSP's classifications of keynote sound, sound signal and soundmark, with their respective forms of background and foreground attention, and the semantic interpretations derived from the "audio language" norms of media, advertising and the accompaniment environments of background music, radio and iPod listening (Truax, 2001).

The broader generalization of soundscape composition to "context-based composition" suggests that "real-world contexts inform the design and composition of aurally based work at every level, that is, in the materials, their organization and ultimately the work's placement within cultural contexts. Perhaps most significantly, listeners are encouraged to bring their knowledge of real-world contexts into their participation with these works" (Truax 2017b, p. 1). This clearly indicates a move toward inclusivity and a removal of barriers between art and design, the everyday and the imaginary, music and environment. In terms of real-world concerns, some of the most pressing today are ecological issues, where sound artists and others wonder what they might effectively contribute. One approach, directly related to acoustic ecology, is to communicate aspects of current environmental crises with sound, such as David Monacchi's Fragments of

Extinction project that is focused on species decline in tropical forests (Monacchi 2016). Jonathan Gilmurray (2017) has outlined the range of current approaches, and grouped them around five categories:

1. Enacting metaphors which facilitate a personal connection with environmental issues.
2. Articulating the harmonious co-existence of humans, technology and the natural world.
3. Allowing us to experience normally inaccessible aspects of the environment.
4. Communicating environmental data through sound.
5. Facilitating community engagement with ecological issues.

After reviewing how sound art might relate to current ecological thinking to create a "re-listening" to the world, or a "rewilding of the ear" (Pinto, 2017), Gilmurray finds "sound art to be an inherently ecological medium—one which enables a direct, sensorial engagement with the agency of vibrant matter, facilitating an embodied understanding of our embeddedness within an inter-connected ecosystem, and creating sonic possible worlds which allow us to re-imagine how things could be" (Gilmurray, 2017, p. 40).

Conclusion

In this chapter I have tried to show how the pioneering work of R. Murray Schafer and the WSP have laid a foundation for the contemporary understanding of acoustic ecology and how it supports current work in sound studies and context-based soundscape composition. Schafer's turn from an anti-noise stance to a listener-centered approach based on the soundscape concept was the critical shift that allowed a subjective, cognitive, and culturally based model to be formulated that described the soundscape as a system of auditory relationships. Enlarging the scope of the model from the individual listener to larger social structures resulted in the concept of the acoustic community, and developed fieldwork methodology to establish how in practice such communities function, and how they continue to evolve, as well as to what forces they are vulnerable. Examples of bioacoustic habitats, as typified by the acoustic niche hypothesis, continue to be inspiring, even as they are currently threatened. And, as with natural habitats, questions about acoustic sustainability and quality of life confront us in the modern world—problems that hopefully the ecological model can begin to address and redress. The research and practices of sound studies and soundscape composition can deepen our understanding and experience of both acoustic and technologically mediated environments, as well as providing alternative sonic worlds we may find inspiring and possibly therapeutic. The wide-ranging work covered here provides many insights into where sound can lead us, as well as how deeply it implicates us in our world.

Note

1. To know more about "acoustic ecology" or "soundscape ecology," as it is sometimes referred to in a scientific ecology context, see Truax and Barrett (2011).

References

Bijsterveld, K. (2008). *Mechanical Sound: Technology, Culture, and Public Problems of Noise in the Twentieth Century*. Cambridge, MA: The MIT Press.

Bijsterveld, K. (Ed.). (2013). *Soundscapes of the Urban Past: Staged Sound as Mediated Cultural Heritage*. Bielefeld, Germany: Transcript.

Birdsall, C. (2012). *Nazi Soundscapes: Sound, Technology and Urban Space in Germany, 1933–1945*. Amsterdam: Amsterdam University Press.

Black, C. (2014). International Perspectives on the Historic Intersections of Electroacoustic Music and the Radio Medium. *Organised Sound, 19*(2), 182–191.

Born, G. (Ed.). (2013). *Music, Sound and Space: Transformation of Public and Private Experience*. Cambridge/New York: Cambridge University Press.

Bregman, A. (1990). *Auditory Scene Analysis*. Cambridge, MA: MIT Press.

Burtner, M. (2011). EcoSono: Adventures in Interactive Ecoacoustics in the World. *Organised Sound, 16*(3), 234–244.

Burtner, M. (2017). Sounding Art Climate Change. In M. Cobussen, V. Meelberg, & B. Truax (Eds.), *The Routledge Companion to Sounding Art*. New York: Routledge.

Cameron, L., & Rogalsky, M. (2017). A Day in Algonquin Park: William W. H. Gunn and the Circadian Audio Portrait. *Organised Sound, 22*(2), 206–216.

Campbell, J. (1983). Ambient Stressors. *Environment & Behavior, 15*(3), 355–380.

Corbin, A. (1998). *Village Bells: Sound and Meaning in the Nineteenth-Century French Countryside*. New York: Columbia University Press.

DeNora, T. (2000). *Music in Everyday Life*. Cambridge/New York: Cambridge University Press.

Drever, J. L. (2002). Soundscape Composition: The Convergence of Ethnography and Acousmatic Music. *Organised Sound, 7*(1), 21–27.

Drever, J. L. (2017). Field Recording Centered Compositional Practices. In M. Cobussen, V. Meelberg, & B. Truax (Eds.), *The Routledge Companion to Sounding Art*. New York: Routledge.

Emmerson, S. (1986). The Relation of Language to Materials. In S. Emmerson (Ed.), *The Language of Electroacoustic Music*. London: Macmillan.

Gilmurray, J. (2017). Ecological Sound Art: Steps Towards a New Field. *Organised Sound, 22*(1), 32–41.

Ihde, D. (1976). *Listening and Voice*. Athens, OH: Ohio University Press.

Ihde, D. (2007). *Listening and Voice: Phenomenologies of Sound*. Albany, NY: State University of New York Press.

Järviluoma, H. (2017). The Art and Science of Sensory Memory Walking. In M. Cobussen, V. Meelberg, & B. Truax (Eds.), *The Routledge Companion to Sounding Art*. New York: Routledge.

Järviluoma, H., Truax, B., Uimonen, H., Kytö, M., & Vikman, N. (Eds.). (2009). *Acoustic Environments in Change*. Tampere, Finland: University of Joensuu.

Kelman, A. (2010). Rethinking the Soundscape: A Critical Genealogy of a Key Term in Sound Studies. *Senses and Society, 5*(2), 212–234.

Krause, B. (1987). Bioacoustics: Habitat Ambience in Ecological Balance. *Whole Earth Review, 57*, 14–18.

Krause, B. (2012). *The Great Animal Orchestra: Finding the Origins of Music in the World's Wild Places*. New York: Little Brown.

McCartney, A. (2016). Ethical Questions About Working with Soundscapes. *Organised Sound, 21*(2), 160–165.

Monacchi, D. (2016). A Philosophy of Eco-Acoustics in the Interdisciplinary Project "Fragments of Extinction". In F. Bianchi & V. J. Manzo (Eds.), *Environmental Sound Artists: In Their Own Words*. New York: Oxford University Press.

Norman, K. (2012). Listening Together, Making Place. *Organised Sound, 17*(3), 257–267.

Pinch, T., & Bijsterveld, K. (Eds.). (2012). *The Oxford Handbook of Sound Studies*. New York: Oxford University Press.

Pinto, A. G. (2017). Towards a Rewilding of the Ear. *Organised Sound, 22*(1), 51–60.

Polli, A. (2016). Sonification of Global Environmental Data. In F. Bianchi & V. J. Manzo (Eds.), *Environmental Sound Artists: In Their Own Words*. New York: Oxford.

Polli, A. (2017). Soundwalking, Sonification, and Activism. In M. Cobussen, V. Meelberg, & B. Truax (Eds.), *The Routledge Companion to Sounding Art*. New York: Routledge.

Schafer, R. M. (1969). *The New Soundscape*. Vienna, Austria: Universal Edition.

Schafer, R. M. (1970). *The Book of Noise*. Wellington, New Zealand: Price Milburn.
Schafer, R. M. (1973). *The Music of the Environment*. Vienna, Austria: Universal Edition.
Schafer, R. M. (1977). *The Tuning of the World*. New York, NY: Knopf. Reprinted as *The Soundscape: Our Sonic Environment and the Tuning of the World*. Rochester, VT: Destiny Books, 1994.
Schwartz, H. (2011). *Making Noise: From Babel to the Big Bang & Beyond*. New York: Zone Books.
Schwartz, T. (1973). *The Responsive Chord*. Garden City, NY: Anchor Books.
Smith, C. (1993). The Acoustic Experience of Place. *Proceedings, The Tuning of the World Conference*, vol. 2. Banff, AB.
Soundscape, The Journal of Acoustic Ecology, wfae.net.
Sterne, J. (2003). *The Audible Past: Cultural Origins of Sound Production*. Durham, NC: Duke University Press.
Sterne, J. (Ed.). (2012a). *The Sound Studies Reader*. New York: Routledge.
Sterne, J. (2012b). *MP3: The Meaning of a Format*. Durham, NC: Duke University Press.
Thompson, E. (2002). *The Soundscape of Modernity: Architectural Acoustics and the Culture of Listening in America, 1900–1933*. Cambridge, MA: The MIT Press.
Truax, B. (Ed.). (1978). *Handbook for Acoustic Ecology*. In R. M. Schafer (Series Ed.), *No. 5, The Music of the Environment Series*. Vancouver, Canada: ARC Publications.
Truax, B. (1984). *Acoustic Communication* (2nd ed.). Westport, CT: Ablex Publishing, 2001.
Truax, B. (1996). Soundscape, Acoustic Communication and Environmental Sound Composition. *Contemporary Music Review, 15*(1), 49–65.
Truax, B. (Ed.). (1999). *Handbook for Acoustic Ecology* [CD-ROM], Cambridge Street Publishing. www.sfu.ca/~truax/handbook.html
Truax, B. (2002). Techniques and Genres of Soundscape Composition as Developed at Simon Fraser University. *Organised Sound, 7*(1), 5–14.
Truax, B. (2008). Soundscape Composition as Global Music: Electroacoustic Music as Soundscape. *Organised Sound, 13*(2), 103–109.
Truax, B. (2017a). Acoustic Space, Community and Virtual Soundscapes. In M. Cobussen, V. Meelberg, & B. Truax (Eds.), *The Routledge Companion to Sounding Art*. New York: Routledge.
Truax, B. (2017b). Editorial. *Organised Sound, 22*(1), 1–3.
Truax, B., & Barrett, G. (2011). Preface: Soundscape in a Context of Acoustic and Landscape Ecology. *Landscape Ecology, 26*(9), 1201–1207.
Westerkamp, H. (2002). Linking Soundscape Composition and Acoustic Ecology. *Organised Sound, 7*(1), 51–56.
World Soundscape Project. *The Music of the Environment Series* (R. M. Schafer, Ed.). Vancouver, Canada: ARC Publications.

(1973) No. 1, *The Music of the Environment.*
(1978) No. 2, *The Vancouver Soundscape.* Originally self-published by the WSP in 1973.
(1977a) No. 3, *European Sound Diary.*
(1977b) No. 4, *Five Village Soundscapes.*
(1978) No. 5, *Handbook for Acoustic Ecology* (Barry Truax, Ed.).
WSP Database. www.sfu.ca/sonic-studio/WSPDatabase

CHAPTER 3

The Disruptive Nature of Listening: Today, Yesterday, Tomorrow

Hildegard Westerkamp

Introduction: Today[1]

When I speak of the disruptive nature of listening, I agree with Michael Stocker (2013) who writes: "Our experience with sound unfolds as a continuous now" (p. xiii). If we open our ears to this experience of sound unfolding as a *continuous now*, it inevitably includes an opening to surprises, to the unexpected, to the difficult and uncomfortable, to noise or potential discomforts with silence. It means staying with the sensations for a time no matter what reactions it may elicit in us.

No doubt we all have had to grapple with discomfort when exposed to disturbing soundscapes or unsettling inner chatter. At such times, do we decide whether we open our listening *further* to the reality of that discomfort and try to affect changes—which is what I would call the disruptive nature of listening—or do we try to ignore it and psychologically shut it out—which is when the sound itself is in danger of disrupting our lives, stressing us, precisely because we are trying to shut out something that our ears and bodies are still receiving, still *per*ceiving? When external pressures are present and we cannot escape them for whatever reason, our focus *has* to turn inside and with extra strength and determination. Inner

H. Westerkamp (✉)
Independent Artist and Researcher, Vancouver, BC, Canada

M. Droumeva, R. Jordan (eds.), *Sound, Media, Ecology*,
Palgrave Studies in Audio-Visual Culture,
https://doi.org/10.1007/978-3-030-16569-7_3

focus, inner listening and expansion become a necessity for survival in such instances.

I had experienced this kind of listening stance consciously for the first time in India in a meditation room, while outside loud car motors and trucks were struggling up a mountain road, or in a temple with people worshipping while crowds and noises, hustle and bustle happened around them. Strangely, there was a feeling of balance in these situations: a strong sense of inner calm and silence in the face of unpredictable sound mayhem from outside. In this culture, I found the inner world to be a source of sacredness that empowers daily life. It is that which seems to give such situations an elevated sense of focus and balance.

None of this means that we should accept unbearable soundscapes in our lives. It simply points out the many complexities we face in the acoustic environment and in the relationship we have to it. Each soundscape situation has its very own characteristics, each culture, each listener reacts uniquely. And it is precisely this complexity that invites one to establish or keep up an ongoing listening practice. In acoustic ecology, it would imply the recognition and acceptance of the continuously shifting sound dynamics that are characteristic of most soundscapes. The shifts occur over time and that fact alone demands continuous openness and flexibility in aural perception from us, no matter what is studied. What is stable and not shifting is the commitment to listening. Such an experience means that we learn to understand how we listen and that the act of listening grounds us within our own inner world from which—in turn—inspiration springs. Listening cannot be forced. Quite the opposite: true receptive listening comes from an inner place of non-threat, support and safety. Paradoxically, while a grounded and calm state of mind, a sense of safety, peace and relaxation are essential for inspiring perceptual wakefulness and a willingness and desire to open our ears, normal routines, habits and patterns will be disrupted and laid bare in such a process of listening; noises and discomforts inevitably will be noticed, and all kinds of experiences will be stirred and uncovered. Listening in fact implies a preparedness to meet the unpredictable and unplanned, to welcome the unwelcome. As such, listening is inherently disruptive as it puts a wrench into habitual flows of time, habitual behaviour of daily life.

From that stance, we may be better equipped to deal with the environmental, social and political challenges of this decade. The intense corporate push for us to succumb to a 24/7 perception of time is threatening to disrupt deep and profound life rhythms in us and as an extension is also disrupting dangerously the earth's biological, environmental and ecological conditions. The following words by Jonathan Crary (2013) might reinforce why we urgently need creative and forceful tools to challenge this trend.

> 24/7 steadily undermines distinctions between day and night, between light and dark, and between action and repose. It is a zone of insensibility, of amnesia, of what defeats the possibility of experience.
>
> ...The planet becomes reimagined as a non-stop work site or an always open shopping mall of infinite choices, tasks, selections, and digressions. Sleeplessness is the state in which producing, consuming, and discarding occur without pause, hastening the exhaustion of life and the depletion of resources. (p. 17)

An ongoing listening practice tends to become a way of life that inevitably reveals and amplifies that which is ignored or normally avoided. Listening not only grounds us within our own inner world from which inspiration springs, but most importantly, it inspires new ideas, new approaches to studying the soundscape, and it changes the quality of soundmaking, speaking and musical expression. Taking the time to listen goes against today's 24/7 status quo of a hectic pace and stress, of racing toward riches and success, of never having time and always being importantly busy. In this larger context, listening is a conscious practice in learning to change our pace in a society dangerously speeding out of control. Out of that *doing* comes an entirely new experiential knowledge. This is precisely what makes out the disruptive nature of listening. The decision to do it, rather than talk about it, is the first disruption of old patterns and gives us a visceral understanding of what in fact the disruptive nature of listening really is.

Yesterday: Beginnings with the World Soundscape Project

The preceding paragraphs were written in 2015 in a 'tone of voice' that most resonates with today's readers, or so it is hoped. As such, it undoubtedly bears witness to the atmosphere of our times, the concerns, preoccupations and ideas regarding the state of the world right now and how we perceive our future. From this platform, I will now attempt to shed light on the 'tone of voice' of the 1970s, when my journey into sound and listening started, when realistically speaking, my career began, working with

Canadian composer R. Murray Schafer and the World Soundscape Project (WSP).

Given this involvement with the WSP and the powerful influence Schafer's ideas and work had on me then, given my personal understanding of the times, my 'being there,' the ways we worked, and given that this work—more than 45 years later—has been debated and critiqued in a variety of ways for a while now, I find myself reflecting on how changing times shed different light on past work, no matter how 'right' and inspiring it seemed in the context and atmosphere of its own time, with its characteristic cultural, social, political and environmental conditions. So much of it seems to have to do with timing, being in the 'right' place at the 'right' time. And that in itself leaves powerful traces in those who were there. By way of describing this atmosphere, let me give you some insight through my own journey. It will be a specific and personal perspective, much of which I shared with the other members of the WSP, but much of it I did not. And those differences of course eventually led to different futures and engagements in each of our lives.

In 1970 or 1971, my habitual patterns of listening were shaken up profoundly when I heard a guest lecture by Schafer at the University of British Columbia (UBC), where I was studying music. It was an AHA moment whose significance was revealed gradually over years to come. The immediate impact was that my ears opened up freely and suddenly to the sounds of my surroundings, and I was inspired and excited to notice them. Later, I recognized that I had experienced a first kind of *ear cleaning*.

Something similar happened around the same time, when I heard a performance of John Cage's *4′33″*. By all indications, we were going to hear a piano piece. After all, it was announced as the last piece in an hour-long piano recital performed by Frances Mary Adaskin at the UBC recital hall. When that expectation was not fulfilled, a gradual shift in listening brought the other sounds in the hall to the forefront of our perception: the small movements made by people in the audience, breathing, bodies shifting, clearing throats, air conditioning, voices from outside, building creaks and so on. The piece had not only redirected our listening attention to other sounds that are normally blocked out in concerts or declared 'disturbing' in the classical music concert tradition, but it also gave us new understanding about the inner workings of aural perception. The quality of the silence in the room was determined by how we listened to the sounds that 'broke' it.

Equally profound was a third experience around the same time that shifted my listening perception and meanings of music and performance. Pauline Oliveros was slated to give a concert at the Western Front in Vancouver. Even though there was nothing conventional about this new artist-run centre, there was still an expectation that the audience would be witness to a music performance. Instead we entered an empty room and were invited to make ourselves comfortable on the floor. Pauline then—sitting cross legged in the centre—proceeded to introduce us to her then-brand-new concept of *Sonic Meditations* and gave us clear and simple instructions about the sounds we were to make with our breaths and voices and how to structure the timeline of this soundmaking. We had become the performers. She was the listener. It 'blew my mind,' as the expression went at the time, and some audience members were outraged that they had been asked to pay for tickets when they themselves in fact had been the performers!

Similarly, Schafer had surprised us by structuring his lecture against our expectations of what a lecture should be. He had placed three or four music stands in different areas of the stage and each one of them was dedicated to a certain topic of music or sound: one to his journey and sound experiences in Persia; the second one to music composition; the third one to soundscape and noise issues. A fourth one may have been dedicated to silence, to not speaking. Throughout the lecture, he moved freely between these topics, connecting them with the sound of his footsteps moving from music stand to music stand (Westerkamp, 2011, pp. 10–19).

As if this was not enough, someone in the audience stood up in seemingly random intervals, interrupting whatever was going on in the lecture and asked questions like *how many airplanes have you heard today* or *what was the first sound you heard today* or *how many birds have you heard today* and more. Of course these individuals—some of them my later colleagues in the WSP—had been placed in the audience intentionally and were instructed to speak at certain points in the presentation. We were in fact listening to a highly composed talk, something I had never experienced before.[2] The unusual format alone, breaking all conventions of standard university lectures at that time, heightened our listening attention and created alertness and utter delight in many of us. And for me personally, my listening woke up from the conventional music studies with which I had grown up and in which I was immersed at that time at UBC.

The moment of leaving the music building after the lecture is imprinted in my memory like a physical sensation: whatever had blocked my listening perception up to that point had been removed completely, as if pulverized. Suddenly and literally, I heard all sounds around me without any mental constructs obstructing my full and welcoming aural attention. I experienced this process as inspiration. In fact, it was as if my ears would never close again, and indeed, this very moment—so I realized later—was the beginning of my life's work.

I have described this experience in some detail here because subsequently throughout my career and particularly in the context of soundwalks, I have witnessed this opening of listening perception, this unplugging of ears, again and again in many people. Enthusiasm and inspiration invariably mark such moments of true perceptual opening. They are a pleasure and often unforgettable, and are in high contrast to so many situations in which we experience rigid approaches to education and work, and when pressure, competition, and fear drives these processes. Open listening is not safe in those contexts and ends up becoming strategic, to 'get through' a job, curriculum or examination.

Schafer's lecture, an invitation to open our ears to the *whole* world, created a sense of liberation and delight. It gave me permission and therefore a sense of security that it was quite okay to apply my listening to more than the tasks of my musical studies. In fact, it placed my musical education into a larger cultural, environmental context and gave it relevance beyond the walls of the music school and its practice rooms. Ultimately it gave me hope that my love of listening, my *ways* of listening, could also find a place of action, work and creation in the world.

After receiving my Bachelor of Music in 1972, I got a job as music teacher in a private school. Almost immediately, I ordered Schafer's *The Book of Noise* (1970) for grades 7 and 8, with the aim to listen to and study the sounds of the Vancouver environment. This caused great controversy among the private school parents, and its relevance in the school's music classes was questioned severely. On the other hand—as I heard later—my book order of 60 copies caused curiosity among the members of the WSP who were wondering who this music teacher might be, integrating the study of noise and the soundscape into the music curriculum of a private school.

It was this experimental approach to music teaching in conjunction with the memory of Schafer's lecture that caused me to contact him spontaneously, one day in 1973. The result was that I began working with the WSP almost immediately. At that time, the group consisted of R. Murray Schafer, Barry Truax, Bruce Davis, Peter Huse and Howard Broomfield.

That I was the only woman did not concern me, although I was surprised by the initial attempts of some of my colleagues to relegate me to jobs at the typewriter and the photocopier. I was passionate about the research work. This and a basically good relationship with my colleagues enabled me to nip in the bud these unconscious assumptions about the work given to a woman. What really mattered was that this new job exposed me to the experience of listening like no other context had done before. The *listening-in-itself* was what catapulted me into a new and exciting way of perceiving the world. Listening as the main perceptual stance toward the environment was the ground on which all our research, writing, recording, as well as the production of books, documents, radio programmes and sound pieces were developed, built and created.

Schafer was writing his seminal book *The Tuning of the World* (1977) at that time, and in this context, we studied and researched just about every aspect of sound that one can imagine and brought it in relation to the book's main focus: the study of the soundscape. Most of us were composers and/or musicians. In the context of our soundscape work, the composer was perceived not only as the designer of musical sound in a composition but also and most importantly as an acoustic designer of daily life. We thought of the composer as a potentially valuable contributor toward dealing with issues of soundscape. Composers—whom one could consider to be the true specialists in listening and soundmaking—could become the socially conscious, sonic architects or acoustic designers of our cities, buildings and villages. It was precisely this—the vision of the artist/composer as a crafts person, as someone trained in *all* disciplines of sound—that attracted me to the WSP (Westerkamp, 2002).

All of us were naturally inclined to use our listening perception as a primary way to relate to the world, and we all wanted to affect changes in the sound environment, but none of us were single-minded about music and composing. Each one of us brought other interests, passions and skills into the group. Howard Broomfield who tragically died during the time of Expo '86 in Vancouver had studied sociology, was a passionate radio listener and later applied this radio sense to a programme on Vancouver Co-operative Radio entitled *The Listener*. His sociologist ear was drawn to people and community soundscapes, in particular, to the disenfranchised voices of society. I remember Bruce Davis as a most natural listener who found his focus in recording the soundscape and was always full of new ideas—such as "Wilderness Radio" (1975), which he developed into a most interesting article and whose central focus recently echoed in the

"Wetlands Project" broadcast, also on Vancouver Co-operative Radio.[3] Peter Huse was also a writer and poet, and at heart a philosopher and thinker, always listening, always reviewing and questioning what we were doing. He, Bruce and Howard were our recordists. We owe them the high-quality recordings made in Vancouver, across Canada and in Europe, that make up the WSP environmental sound recordings archive now. Barry Truax had studied science as an undergraduate and was deeply involved in developing computer music technology and programming. Among all of us, he was most equipped to edit documents such as the *Handbook for Acoustic Ecology* (Truax, 1999) and later to take over Schafer's position at SFU. I discovered a love for research, while my other interests were located in education and environmental/political issues and my activist self was woken up intensely. Only later did I discover a passion for recording the soundscape and composing with environmental sound.

In short, each of us brought a certain emphasis of interests to the group, which gave it a multidisciplinary atmosphere—something that was rather new at the time. We all were interested in bridging disciplines, gaining knowledge from other fields and bringing it into the research focus of our work, which we called *soundscape studies* at the time. And Murray, unquestionably our guiding hand at the time, keeping the group focused, was welcoming our often very individualistic, even eccentric contributions. We spent hours in the Sonic Research Studio listening to the field recordings that had just been made. Valuable time was spent reflecting on and understanding sound behaviour and our own perception. Our listening was noticed, respected and allowed to flourish. Critical thinking, personal impressions and outspoken opinions were invited. There was a sense of urgency in our soundscape activism: we pushed and reached for a deeper socio-cultural awareness. We began to understand that ecological imbalances in the soundscape do not occur separately from us but are intricately intertwined with our relationship to the environment and how we listen and make sound.

Schafer envisioned the WSP as a type of umbrella for all disciplines concerned with sound, noise, acoustics and music. The profoundly enriching experience of working in this group grew from our effort to combine scientific knowledge with artistic, perceptual knowledge of sound. It was precisely in this combination of research, education, creativity and activism in which the energy of the WSP was located and which enabled us to produce in a relatively short time a number of pioneering documents and projects. But at a time when multidisciplinary approaches were relatively new and still viewed with suspicion (and often labelled as unprofessional, even flaky), it was a challenge to convince specialized professionals to join us in our soundscape research.[4]

Yesterday: The "Tone" of the 1970s

Let me return to the concept of the disruptive nature of listening as discussed in the first part of this chapter. The highly inspired atmosphere within the WSP to keep listening to everything and learning in order to deepen our understanding of the world was what created a sense of safety and containment in us. We were never forced to listen, we were *encouraged* to listen and, more importantly, to believe in our ears and in our very own ways of perceiving the environment. This was a profound teaching at that time and a disruption of our habits, of what we were 'supposed' to hear. The idea of listening to environmental sound and noise disrupted not only *our* habits but most significantly also the traditional approaches to the noise problem: most noise activists did not know what to do with it and certainly did not want to slow down for soundwalks; and for scientists it put a wrench into the lab approach of psychoacoustic studies, noise measurements and so on. In that sense, it resonated with a growing trend in the 1960s and 1970s to revolutionize and question social, political and cultural patterns in Western European and North American countries. Vancouver was partaking intensely in this trend and, as a result, was an exciting place for us 20-somethings to test our socio-cultural curiosities and experiment with new content, formats, structures and media.

It was within this context of questioning and challenging every possible convention that Schafer's book *The Tuning of the World* originated. As a relatively new immigrant from Germany, I experienced this as a profound awakening, filled with opportunities to change a world that was still healing from the deep scars of WWII and Nazi Germany. I don't think we can underestimate the powerful atmosphere of horror, guilt, puzzlement and confusion that was still hovering underneath the flashy economic growth of the 1960s and 1970s in Europe and North America and that ultimately caused us, mostly the post-war generation, to break out into protests and revolutionary actions, aiming to push through old institutional brick walls, breaking cultural formats and traditions that could no longer be trusted. We *had* to try out new ideas.

We were also highly influenced at that time by the protests in the US against the Vietnam War and felt the results of it in the influx of draft dodgers who were entering Canada in droves and hiding out in various parts of Interior British Columbia. Student revolts that were happening in Berkeley, California, were echoed at UBC and at SFU, the Haight-Ashbury hippy scene in San Francisco was mirrored on Fourth Avenue in Vancouver. The Pacific West coast seemed at the far edge of European and North American civilization, and Vancouver was still far from being a gateway to the Pacific Rim. Culturally, this city was wide open and full of possibilities, seemingly unrestricted by any conventions and traditions.

When I first arrived from Germany, I had felt a great sense of relief that I was far away from the deeply entrenched standards of European cultural traditions, with which I had grown up. Anything was possible here, and no one was going to judge, put down, or curb what we wanted to explore with our creative energy. This context felt safe and therefore encouraged us to open our senses in new ways. That there was a profound cultural and social silence of indigenous voices and expressions, or to say it another way, that the indigenous presence had been relegated to specific places and contexts in the Canadian societal landscape, only dawned on me gradually when I was not such a new immigrant anymore. It was quite shocking to discover, how long it took me—this immigrant of European descent, who thought she was so progressive—to open my ears, my entire being, to this cultural presence that had been silenced for too long.

Preoccupied with my European past and with the desire to revolutionize the world, I did not yet hear the significance of that silence. The other members of the WSP, who all were born in Canada or the US, heard it more perhaps, and to varying degrees, but like almost everyone at that time, only to a limited extent. The nature of that silence—the suffering, pain and struggles it contains—is finally being revealed more and more, and the conditions for open and compassionate listening that can lead to real changes seem to be more promising nowadays. It is precisely because of this gradual shift in societal consciousness that Randolph Jordan can now—in his 2015 text below—point out a significant weakness in the WSP's work, of which we were not entirely aware at the time:

> On the issue of cultural politics we find less progress across the two official Vancouver releases.[5] Both make some mention of indigenous presence in the land and tie these references to their concerns over urban noise pollution. The '73 LP opens with the sound of the ocean primordial, lapping against the shore, with wind and birds following in short order. Amidst these, the voice of a Squamish man begins speaking in his native tongue, soon interrupted by an

> emerging seaplane flying low overhead, ushering in the era of Vancouver's incorporation as a city and the ensuing industrial development. Only someone who understands the spoken language here will know if the WSP's narrative runs counter to the story being told by the Squamish man, for no translation is offered either in the grooves or within the jacket. (para. 9)

He is correct that there is no exact translation of Herbert George's words, but both releases state the theme of his narrative in *The Vancouver Soundscape* (WSP, 1973) liner notes—*He tells us how he makes racing canoes out of 60-foot cedar logs*—and also clarify how this section was put together and contextualized. I remember that we prided ourselves at the time of offering transparency to the listener, wanting to demystify the editing or mixing process, in order to make a clear distinction between listening directly to the environment and listening to recordings of soundscapes. This distinction became particularly poignant at the time, when radio listeners and TV viewers increasingly became aware and suspicious of the kind of manipulation and subliminal messaging that occurred in media advertising.

The fact that we were now able to make high-quality environmental sound recordings in the midst of an intensifying commercial consumer atmosphere compelled us to be transparent about such details in our documentation. We had no models for this kind of work. No one had attempted opening ears to an entire city before, let alone trying to give a comprehensive sonic impression of such a large urban expanse.[6] So, to get back to Randolph's words above: the fact that a seaplane disrupts the quiet soundscape and Herbert George's narrative, was in fact a deliberate sound design decision. From our soundscape awareness stance, it highlighted the fragility of human language, of quiet and silence, and how easily these sounds and soundscapes can be masked or destroyed by such powerful sounds as aeroplane motors.

The way we point out in the liner notes that Herbert George has forgotten many of the words could nowadays be read as culturally insensitive to indigenous reconciliation, perhaps as a racial slur.

> The language is his mother's tongue, Squamish; but his narrative is halting because he has forgotten many of the words. During the pause (at 4:18) you may hear him mutter, "How do you say that now?" Herbert George uses white man's tools and their names appear conspicuously in his narrative (spirit level, square, bit, plane, sander, marine varnish). (WSP, 1973)

But in the 1970s, it was not yet common knowledge that all Indian children had been forced to stop speaking their indigenous languages as soon as they were taken away from their home environments into residential schools. This would have been a good opportunity indeed to elaborate on those issues, if we had been better informed and more aware at that time. Let me stay with Randolph's (2015) discussion of the WSP's lack of engagement with the presence of First Nations in Vancouver society. He goes on to state:

> Similarly, in a brief documentary about Vancouver's changing sounds on the '96 release, Westerkamp comments on the intrusive presence of the air conditioning system in the Museum of Anthropology on the U.B.C. campus, marring visitors' experience of "some of B.C.'s most fascinating Native artifacts." Westerkamp's respect for indigenous culture comes through clearly enough. Yet the WSP's presentation of local indigeneity takes a step backward here as the living voice speaking a dying language heard in '73 has given way to just plain old dead people in '96, best contemplated in the quietude of Arthur Erickson's high modernist design for the building. (para. 11)

What Randolph could not know was that this one sentence referring to 'Native artifacts' was based on a rather critical article I had written in 1976/1980 about the soundscape of the interior of the then "new" Museum of Anthropology, designed by Arthur Erickson, with the rather provocative title, "The New Museum of Anthropology in Vancouver: An Acoustic Dump." The initial enthusiastic response from *The Canadian Architect* indicating in a formal letter that they "would like to publish the article,"[7] and setting the logistics of the publication process in motion, was eventually followed by complete and utter silence, even after I inquired several times about the status of my article. The article had ruffled quite a few feathers, including those of Arthur Erickson himself. However, he wrote a relatively positive and responsive letter to me, starting with these words: "I was interested in your report… and although it alienated me a little when you were trying to make your point too strongly, I feel that your case is well justified and appreciate the suggestions you make." He even goes on to refer me to the people responsible for some of the exhibits in the museum in order to discuss possibilities for improvement of the soundscape.[8]

On the same day, he wrote a letter to Murray Schafer—they knew each other and had been communicating about issues of sound in architecture over the years—in which his man-to-man tone expresses the same issues slightly differently: "I think it professionally shocked me, if that is what you are hoping for! I was a little disturbed by the aura of suppressed hysteria about Twentieth Century buildings but very much drawn to the possibility of recorded sound such as sea sounds and bird sounds to heighten the presentation of the exhibits." And Schafer responded, "You are brave indeed to have weathered Hildegard Westerkamp's article, though I know she wrote it with passionate concern for improving the new Museum."[9] At least he had the decency to send me a copy of this exchange and no doubt thought that it was all pretty amusing!

The sharpest criticism came from the Museum Director at that time, who reacted particularly outraged to the end of my article entitled "The Totem Pole as an Acoustic Event" (Westerkamp, 1976), where I wrote:

> …the poles have been removed from their original context, and therefore an ideal experience is no longer possible. We must ask ourselves, therefore, if under the circumstances, it is still possible to experience them in a meaningful way, to recognize their images, to hear the stories they tell…
>
> Every totem pole tells a story. It is as much a carrier of an aural tradition as are the songs, narratives and rituals of an aural culture. Totem poles stand silently; listening to the forest, ocean, wind and rain, to ravens and frogs and eagles: listening to everything they are telling us about; made of the materials their surroundings are made of, their surrounding collected in them and brought to new expression in their large silent structure. Each totem pole is an acoustic event for those who want to hear, for those who listen to its voices and the stories it has to tell.
>
> Ideally it is for the survival of this "acoustic event" and not just for the physical survival of the totem pole that the architect must aim for in his design of the museum. (para. 29–31)

The museum director at the time vehemently balked at my suggestion that a totem pole could be understood as an acoustic event and that it may speak to us about something universally meaningful. Although he did concede that it was an interesting view, he thought of it as a romanticized Western interpretation of totem poles and indignantly proclaimed that these poles do not and never did speak for themselves.[10] To be fair, communication with the museum continued and became a little more constructive. After I revised my article several times, partially because some small acoustic changes were in fact happening, the museum director invited me to give a presentation "that sensitizes his students to listening to sound." I continued to do several presentations, but ultimately, the basic idea of changing the sound design in the museum was never pursued.

Even though it is obvious that all ways of language expressions are a result of the times in which they were 'intoned,' it may be important to highlight this specifically in the context of acoustic ecology and soundscape studies, before more time is wasted on superficial critiques of Schafer's work and his tendency to rant in his writing. We were convinced—and when I say 'we,' I mean the post-war baby boomer generation—that only shock effect could bring change and rattle the strong conventional barriers we perceived in the 1970s. Hence my provocative and blunt title for the museum article. One cannot underestimate how the context of that time compelled us to push boundaries and rebel against senseless rules. We felt that only provocative language would bring about real change in a deeply wounded time, in which our generation did not want to repeat damaging social, political and cultural patterns of authoritarian societies.

Schafer's writing style and tone, that *wanted* to rattle and wake up, unfortunately but perhaps understandably in our times, has irritated and alienated some academic writers and thinkers enough that his basic message gets lost in the shuffle. It takes a deeper listening not to let the ranting—which the rattled reader perceives as a kind of 'noise'—mask his fundamental message: sound studies and research can only be complete if they include the essential information and knowledge that is gained through perceptual experience.

The irony of course is that at the time, Schafer was hailed as a 'pioneer' when he rattled people's perception, whereas my own such attempts were characterized as having an "aura of suppressed hysteria!" However, this plus Schafer's extremely hurtful behaviour when he fired me for no good reason[11] could not destroy what had been sparked deeply in me and became a guiding light for the rest of my life. Getting fired in such a senseless fashion put a wrench into my life, as it created difficult conditions and unforeseen obstacles to pursue a path that was still new and mysterious to me. All I knew was that I wanted to tread this path of listening to the world, and ultimately to understand the deeper and very real complexities of such a path, that is, creating safe and trusting conditions from which we can truly listen to each other and the environment and thus allow for deeper disruption, deeper understanding of our ideas. That depth, I believe, can bridge time-dependent wordings, tone of voice and so on;

and the main ideas that inspired us from the start can hopefully remain significant for future work in acoustic ecology and soundscape studies.

Tomorrow: New Tone of Voice, New Approaches

Superficial critiques aside,[12] there are hopeful signs indeed that thoughtfully critical interpretations and follow-ups of Schafer's and the WSP's work are creating a new momentum for innovative dealings with ecological issues in our soundscapes. In fact, consciousness of the importance of sound and listening literally has exploded, in many cases quite independently of Schafer's legacy or the ongoing work of the World Forum for Acoustic Ecology (WFAE).[13] There are too many contemporary works to mention here (Ceraso, 2014; Dixon, 2011; McCartney, 2010); safe to say familiar ideas from the 1970s reverberate into a language appropriate for today, for younger ears, and that is invaluable. This in itself is a sign of our times; it is 'in the air,' so to speak, that there is a need, an urgency perhaps, to rebalance modern life with new approaches. It is amply clear to many—whether professionals or not—that acoustic ecology, sound studies, acoustic communication, soundscape ecology, eco-acoustics, deep listening or whatever title we want to apply to the field, that these studies have an important role to play in grappling with the enormity of environmental, social, and cultural issues that our world faces.

It is not surprising to find a similar echo in Peter Cusack's (2017) small jewel of a book *Berlin Sonic Places, A Brief Guide*. There is a quiet integrity to Cusack's approach that seems in tune with what listening really is. His writing encourages listening. Whereas the WSP tried to study the entire Vancouver soundscape in the mid-1970s and attempted to give a comprehensive sonic impression of the city—excited at the time by the novelty of such an idea—his approach highlights the *impossibility* of such an attempt. And it is precisely that which makes this document so comprehensive to today's reader who is invited into the process, encouraged to go out there and listen, to understand his or her own relationship to a place:

> ...while the whole sonic panorama may elude us, what we actually experience is equally profound and significant... When moving around a city we pass through a continual succession of local soundscapes that merge, often unnoticed, from one to the next. Each is connected to a particular place or spot. These are the city's innumerable sonic places – the building blocks of the urban environment. (p. 17)

Epilogue: An Extended Practice of Listening

Just as one cannot transmit in words the experience of listening that occurs on soundwalks, one cannot claim knowledge of life in places far away from urban centres, if one has never lived it. Although hearing stories or reading books, seeing photos or watching films about such a life can touch us, sometimes profoundly, it can only transmit a lifeless inkling, compared to what people know when they have lived there for many years.

To move from a city to a rural community, particularly in a country like Canada, where many human settlements exist far away from urban amenities and conveniences, where access to electricity, water, and food cannot be taken for granted, is not unlike emigrating. One enters an entirely new way of living, a new culture with different priorities. It takes years to adjust, years to settle, but once it happens, the ground under ones' feet seems differently supportive, as our steps feel the soil and its many variations in shape, texture and buoyancy. We discover that the soundscape, more often than not, speaks clearly and directly to us and thus accompanies us differently through the seasons, day and night. Significantly, our own sounds are very present—tend to stare us in the face, so to speak. Very little can mask our movements and behaviours.

Such a move quite naturally and inevitably brings about an awareness that daily life proceeds differently and may be perceived as an act of rebellion in a 24/7 world—resisting what Crary described earlier as "a non-stop work place or a zone of insensibility, of amnesia, of what defeats the possibility of experience" (2013, p. 17).

This brings us to indigenous populations around the world who have always lived far away from urban centres, who—despite having endured the destruction of their environments, languages and cultural expressions for generations—continue to carry deep-seated knowledge of how best to live in a respectful relationship with the natural environment, its seasons, its cycles and rhythms, how to live in an atmosphere of mutual give and take. Are our listening ears truly welcoming, open and compassionate enough to take in the enormity of the damage that has been done—equivalent to the damage that has been and continues to be done to many natural environments? Is our listening open enough to enter into a true relationship with indigenous people, their cultural and environmental knowledge?

Real reconciliation or de-colonization can only happen in an atmosphere of mutual respect and in safe conditions from which we can truly listen to each other. For we must be prepared that such listening is bound to disrupt many ingrained patterns and expectations of how a society should work. It implies a preparedness to hear out the unexpected and unpredictable, which in turn would allow for the emergence of entirely new visions of how we might co-exist peacefully and in inspired ways with each other and our surroundings.

Notes

1. The introduction consists of excerpts from my keynote presentation *The Disruptive Nature of Listening* at ISEA2015, Vancouver, B.C. Canada, August 18, 2015.
2. Of course, John Cage had already written and performed his "Lecture on Nothing," but it was not until some years later that I became aware of it. See: *Silence*, Lectures and Writings by John Cage, Wesleyan University Press, Middleton Connecticut, 1973, p. 109.
3. A celebration of Earth Day with Vancouver Co-operative Radio's *Wetland* broadcast on Sunday, April 22, 2018—a 24-hour experience in "slow radio" created by artists Brady Marks and Mark Timmings. http://wetlandproject.com/ra.php
4. The past three paragraphs are quoted from Westerkamp, Hildegard. "The Golden Ball – A Foreword (of sorts)," in *A Festschrift for R. Murray Schafer on the Occasion of his 80th Birthday*, Sabine Breitsameter and Eric Leonardson (Eds.), Dieburg Series on Acoustic Ecology 3, University of Applied Sciences, Darmstadt, Germany. p. 10.
5. He is referring to the original *Vancouver Soundscape*, consisting of two LPs and a book, published in 1975 and the partial rerelease of the LPs in the format of two CDs with additional soundscape compositions in 1996.
6. Of course, in the process of our research and productions, we became aware that there *were* others researching, recording and exploring soundscapes at the same time we were—such as Tony Schwartz (LP *New York 19*, released by Folkways Records in 1955), Irv Teibel (*environments* LP series, 1969–79), Bernie Krause (*Wild Sanctuary*, dedicated to recording natural environments since 1968), Walter Tilgner (nature recordist since the late 1960s) and more.
7. Letter by Robert Gretton, Managing Editor of the *The Canadian Architect*, to Hildegard Westerkamp, November 26, 1976.
8. Excerpt from letter by Arthur Erickson to Hildegard Westerkamp, July 8, 1977.
9. Excerpt from letter by R. Murray Schafer to Arthur Erickson, August 1, 1977.
10. Letter from Michael M. Ames, then director of the Museum of Anthropology, to Hildegard Westerkamp, December 1976.

11. I was not fired because of faults in my work, but because my enthusiastic and committed contributions to the group's research were deeply misunderstood and inexplicably caused jealousy in the context of Schafer's relatively new marriage! Had I been a young man with the same kind of passion for my work, none of this would ever have happened.
12. Such as the popular critiques that Schafer's acoustic ecology is all about fetishizing wilderness and countryside, which have swung the pendulum the other way and made it very fashionable—as in Labelle and J. Lacey's work—to now glorify city noise.
13. The WFAE was formed during the *First International Conference on Acoustic Ecology* in Banff, Alberta, Canada, on August 13, 1993. See: https://www.wfae.net/

References

Ceraso, S. (2014). (Re)Educating the Senses: Multimodal Listening, Bodily Learning, and the Composition of Sonic Experiences. *College English, 77*(2), 102–123.

Crary, J. (2013). *24/7, Late Capitalism and the Ends of Sleep*. London: Verso.

Cusack, P. (2017). *Berlin Sonic Places, a Brief Guide*. Berlin, Germany: DAAD Artists-in-Berlin Program.

Davis, B. (1975). FM Radio as Observational Access to Wilderness Environments. *Alternatives, 4*(3), 21–27.

Dixon, M. (2011). *Ways of Hearing*. Retrieved from http://www.soundandmusic.org/projects/ways-hearing

McCartney, A. (2010, September 27). *Soundwalking: Creating Moving Environmental Sound Narratives*. Retrieved from https://soundwalkinginteractions.wordpress.com/2010/09/27/soundwalking-creating-moving-environmental-sound-narratives/

Randolph, J. (2015, September 9). Unsettling the World Soundscape Project: The Bell Tower of False Creek, Vancouver. *SoundingOut! Blog*. Retrieved from https://soundstudiesblog.com/2015/09/03/unsettling-the-world-soundscape-project-the-bell-tower-of-false-creek-vancouver/

Schafer, R. M. (1970). *The Book of Noise*. Wellington, New Zealand: Price Milburn.

Schafer, R. M. (1977). *The Tuning of the World*. New York, NY: Alfred A. Knopf. Reprinted as *The Soundscape: Our Sonic Environment and the Tuning of the World*. Rochester, VT: Destiny Books, 1994.

Stocker, M. (2013). *Hear Where We Are, Sound Ecology and Sense of Place*. New York: Springer.

Truax, B. (1999). *Handbook for Acoustic Ecology* (2nd ed.). Vancouver, Canada: ARC Publications.

Westerkamp, H. (1976). The New Museum of Anthropology – An Acoustic Dump. Retrieved from http://hildegardwesterkamp.ca/writings/writingsby/?post_id=23&title=%E2%80%8Bthe-new-museum-of-anthropology-in-vancouver:-an-acoustic-dump

Westerkamp, H. (2002). Bauhaus and Soundscape Studies—Exploring Connections and Differences. In P. Luckner (Ed.), *Anthologie: Multisensuelles Design*. Halle, Germany: Hochschule für Kunst und Design.

Westerkamp, H. (2011, August). Ears Unplugged, Reflections on 40 Years of Soundscape Listening. *Japanese Journal of Music, 9*(1), 10–19.

World Soundscape Project (WSP). (1973). *The Vancouver Soundscape [LP]*. Vancouver, Canada: Simon Fraser University.

CHAPTER 4

Local *Ear*donances: Raymond Murray Schafer's Contribution to the History and Present-Day Practice of Noise Abatement

Karin Bijsterveld

Introduction

A true acoustic ecologist would probably frown upon reading the title of this chapter. Raymond Murray Schafer's contribution to noise abatement? Wasn't that exactly what Schafer and his colleagues in the World Soundscape Project (WSP) were *not* after? Hadn't the negative cause of noise abatement societies and their complaining members been the thing they wanted to leave behind? Ear cleaning, raising awareness of the sonic environment and enriching it with soundscape compositions and acoustic design were the key themes in their work.

But when I recently revisited the websites of the European noise abatement societies whose twentieth-century histories I had previously studied in *Mechanical Sound*, or the successors of those societies in case the original ones had perished, I noticed that several of them had clearly taken up Schafer's vocabulary, goals or tactics. For sure, not all of them had done so—their foci had been left untouched. But not only was the legacy of the World Soundscape Project visible in hundreds of new initiatives mapping

K. Bijsterveld (✉)
Maastricht University, Maastricht, Netherlands
e-mail: k.bijsterveld@maastrichtuniversity.nl

M. Droumeva, R. Jordan (eds.), *Sound, Media, Ecology*,
Palgrave Studies in Audio-Visual Culture,
https://doi.org/10.1007/978-3-030-16569-7_4

urban and natural soundscapes, it was also noticeable in many of the actions traditional noise abatement societies were involved in. Yet how did and does their uptake of Schafer's ideas relate to their more traditional noise abatement strategies? How do they combine the two approaches? And what might be behind it? I will try answering these questions by analyzing the websites and publications of three of the European noise abatement societies I had examined earlier: the UK Noise Abatement Society (NAS), the Dutch Society Against Noise Nuisance (NSG) and the Working Group on Noise of the German Acoustical Society (ALD)—the successor of the German Working Group for Noise Abatement (DAL). I will focus on half a decade: the years between 2011 and 2015.

This chapter will start, however, by looking back on Raymond Murray Schafer's contribution to the history of noise abatement. This should not be read as analyzing how Schafer's key concepts affected historians' work on noise. What I aim to do is both critically and sympathetically discussing one particular aspect of Schafer's historical work: his writings about local ordinances, or local bylaws. When preparing for *The Soundscape: Our Sonic Environment and the Tuning of the World* (1994/1977), Schafer and his students sent out hundreds of questionnaires to cities around the world to learn about their past and present bylaws concerning noise. The answers by civil servants, so I know from studying the WSP archives, varied widely in terms of how detailed they were or how far they went back in history. Some of the city authorities involved had clearly asked their archivists to find whatever was relevant, while other civil servants focused on the present or did not respond at all. Whereas Schafer's questionnaire was not how historians had learned to do research, and not all findings fit into what historians unraveled later, Schafer's approach showed, for one, the wide range of issues covered by these ordinances, and thus their potential.

This chapter will therefore end with a suggestion to harvest the increasing interest of European noise abatement societies in the legacy of Schafer's acoustic ecology by taking his somewhat forgotten work on the diversity of local ordinances seriously. This can be done, so is my idea, by linking Schafer's widely cited ideal of the soniferous garden, if in a new form, to local ordinances—in *ear*donances that entitle urban citizens to the use of acoustically secluded and enriched spaces. My cases in point are today's tensions, notably in European cities, around the sounds that are considered instrumental in making metropoles attractive: those of music festivals. I will make the idea to create *ear*donances more concrete with the help of recent societal clashes around festivals in Berlin and Amsterdam.

Acoustic Design Versus Noise Abatement

In *The Soundscape*, originally published in 1977, it takes Schafer only a few introductory paragraphs to get to the very heart of the matter. To ensure that the world really tackles the issue of noise pollution, noise abatement should not be foregrounded. It is too negative. Instead, we "must seek a way to make environmental acoustics a *positive* study program. Which sounds do we want to preserve, encourage, multiply?" As long as noises "are the sounds we have learned to ignore," a positive program can only start with learning to listen carefully (Schafer, 1977c, p. 4). In a technology-infused world, just reducing or suppressing sources of unwanted sound is an endless uphill battle anyhow. It is a better idea to map the world's entire soundscape and re-orchestrate the sonic environment at specific locations with pleasant sounds by acoustic design. In the book's 13th chapter, simply entitled "Noise," Schafer concedes that he has had quite some difficulty convincing publishers that his intended book project is not on noise pollution in any conventional way. In his view, there are more than enough of such publications, and he has already dealt with the issue himself in *The Book of Noise*.

The Book of Noise, originally published as the first report from the World Soundscape Project, acquires a second life when the New Zealand publishing firm Price Milburn and Company Limited brings it out in 1973. It is a richly and humorously illustrated educational booklet. Its tone, however, is tough. It includes many of the terms and twists that we know from Schafer's later work. The soundscape is "an important part of our environment," polluted by the dump of "sonic sewage" in which our own voices "are in danger of being drowned" (Schafer, 1973, pp. 3–4). Schafer's original take on composing the sonic environment prefigures dominantly. But notably in a gloomy version. Man has been the world's chief as well as most horrible composer, producing an "exasperating concert" (p. 2). By creating ever-multiplying sets of loud sound-producing tools and gadgets—power-mowers, radios, record players, tape recorders, power saws, amplified guitars, television sets, jackhammers, generators, cars, helicopters, jets and supersonic transport—humans have created an apocalyptic pandemonium of noise that kills their hearing and threatens their hearts. Schafer does not even shy away from speaking about a "sound holocaust" (p. 26) to express his disgust of what is audible in today's societies, which no longer have "some built-in philosophy of restraint" as used to exist in the past (p. 22).

He also expresses his wonder about the old-fashioned character of local bylaws in Canada.[1] Citizens might find local bans on the noises produced by hawkers, peddlers or news vendors, but nothing about the sources of sound that bother them so much today. Citizens should listen closely to their own cities, and behave as "perceptive tourist[s]" keeping "sound diaries" when visiting pleasant soundscapes elsewhere (p. 28). But they should also know that engines can be muffled, should volunteer in or establish noise abatement societies, and contact municipal or national authorities to make them aware of noise pollution. "Above all, the next time a noise annoys you, *complain*" (p. 29).

A Questionnaire on Sound Politics for *The Tuning of the World*

The advice to contact authorities to complain about noise is a far cry from Schafer's approach in *The Soundscape*. In the book that will make him famous, Schafer replaces, as said, noise abatement by enhancing citizens' awareness of the world's soundscape through ear cleaning, soundwalks and sound diaries, and by redesigning it through acoustic design. *The Soundscape* still has a chapter on noise, but it is not meant to join the choir of complainants—it is intended to show the cultural diversity of what is and has been considered noise.

To be able to do this, Schafer claims to draw "extensively" on a research project that had focused "on by-laws and antinoise procedures from over two hundred communities around the world." He expresses his gratitude to the "countless municipal officers" who had sent "copious information" in response to the questionnaire that had been part of the examination. He underlines, however, that it has not been the purpose of the questionnaire to develop a model city bylaw, even if "we could probably do that." It is the variation in what culturally counts as noise that he wants to present (Schafer, 1977c, pp. 181–182).

It is interesting to reflect for a moment on the style of this particular WSP project. Although Schafer uses the results of the queries to underpin historical and anthropological claims, the questionnaire's set-up resembles that of social geography questionnaires at that time. It documents local noise abatement by using informants rather than checking out the archives of these municipalities, as historians would have done. Of course, the unprecedented coverage of the questionnaire would have been impossible with a conventional historian's approach. Moreover, the method is fully in line with Schafer's spatial focus, expressed in his interest in *mapping* the soundscape as a landscape, which in turn draws on earlier work by geographer Michael Southworth, as Jonathan Sterne has explained (2013, pp. 186–187). As I already noted, however, Schafer and his peers had to rely entirely on the selection made and detail offered by those to whom city mayors and other authorial persons had delegated answering the questions.

The letter that Schafer and his team members sent out to cities around the world in the summer of 1973, says they are "currently undertaking a world-wide survey of noise pollution" in order to better understand the scope of and potential solutions for the problem. "No modern city is without noise pollution problems today," the letter's opening paragraph states. It then asks the cities to return copies "of all current anti-noise legislation (bylaws etc.) in force in your city," and to give some "indication as to whether this is effectively enforced." More specifically, the WSP team invites the cities to provide the number of complaints and convictions under the heading of the cities' anti-noise legislation obtained in 1970, 1971 and 1972. In addition, they request information on noise legislation that the cities might have in preparation, ask to "indicate which forms of noise constitute the largest problems in your city," and to send any "historical information available from your archives concerning anti-noise legislation from past periods in the growth of the city." Cities are allowed to return the information in their own language—it will be translated afterwards.[2]

And so these cities did—if they responded. Of the 240 European cities contacted with the letter, 60 responded. In some countries, just one or a few cities responded—in Belgium only Liege returned the questionnaire, with little information. Only a small minority of all responding municipal officers had an answer to the question about noise abatement legislation in the past.[3] Most of them claimed that, to their knowledge, such historical data were not available. Others sent copies of bylaws from the 1930s that were officially still in operation, by way of answering the question on current legislation. Most of the letters, however, did not clarify whether the responding officers or their delegates had indeed checked the archives for information, or hadn't invested in doing so. This makes the historical graph plotting the number of major bylaws and amendments that had passed over time on the basis of the historical documents and data provided by the cities questionable (Schafer, 1977c, p. 194). There is little doubt that the 1960s and 1970s saw a growth of noise abatement legislation (Bijsterveld, 2008, pp. 240–247), but the low level of regulation in the 1910s, 1920s and 1930s may be an artifact of the method chosen. There is now ample evidence of noise abatement regulation in those decades (Bijsterveld, 2003, p. 119; Mansell, 2017, p. 59; Radovac, 2012, pp. 292–295; Thompson, 2002, pp. 126–127, pp. 151–152).

In terms of showing the variety in the noises covered by the bylaws, however, the results of the questionnaire are highly valuable. Schafer spends a few lines on the commonalities in bylaws: "shouting or creating a disturbance in public; street and house music; loudspeakers, radios etc.; noisy animals; unmuffled motor vehicles; noisy industries in residential areas" (p. 195). But he shows more interest in the cultural diversity expressed in the local regulations. These range from banning, between particular hours, using shutters in Italy and bowling in Luxembourg, to beating carpets in German-language countries and barking dogs in Northern countries. And the recorded complaints vary from mah-jong parties in Hong Kong and drum-beaters in Mombasa, Kenya, to family reunions in Rabat, Morocco (pp. 197–199).

This local level is worth reconsidering for the type of sonic politics Schafer had in mind in *The Soundscape*. Because of the difficulty of effectively executing measurements as well as the complicating psychological dimension of perceiving nuisance, he considered a combination of quantitative and qualitative provisions the most practical approach. But he expected and feared "an erosion of the qualitative in favor of the quantitative, for the latter better suits the technocratic mind" (p. 197). Clearly, he was not only critical of the industrialized urban life, and of technology, but also of technocratic, one-size-fits-all solutions. Acoustic design tailored to the cultural and spatial specifics of particular locations should be one of the alternatives.

As many scholars have stressed, however, Schafer's approach to city sounds and technology was rife with ambiguities. Schafer followed Southworth, so Jonathan Sterne (2013) claims, in deploring urban noise and fostering the argument "that high contrast between foreground and background sounds makes sonic spaces more identifiable; and that open space and responsive spaces hold special potential for experimentation and staging sonic events." (p. 186) But Schafer was more fundamentally critical of urban life than Southworth, Sterne adds, as he considered smaller social units, and thus villages, less alienating than the bigger ones in cities. At the same time, Schafer documented both the sounds of villages (Schafer 1977a) and cities (1977b), and approached these sounds as forms of music to which he aimed to open citizens' ears, like John Cage had done before him (Schafer, 1977c, p. 5, 111). And the recordings in which he brought these sounds together, so Ari Y. Kelman has underlined, used the very sound technologies that Schafer attributed to cause "schizophonia," the artificial separation of sound from its original source and place (Kelman, 2010, p. 218, p. 228).

But I depart from Kelman's claim that Schafer had nothing to offer to those suffering from the modern din but disciplined listening and mental orchestration (Kelman, 2010, pp. 217–218). For one, Schafer's stance on bylaws was more consistent with his overall views than his take on technology. A good bylaw had to effectively deal with "the most displeasing sounds of a given locale" (p. 182). What he was squarely against, however, was just banning particular sounds, merely putting these on "the proscription list" as "we do it the ultimate honor of making it all-powerful" (p. 202). The final power, in Schafer's world view, was, and had to be silence. In well-designed parks, or soniferous gardens, nature should be allowed "to speak for itself," whereas "electroacoustic tricks" should be left out (Schafer, 1977c, p. 247). Staging the rich sounds of water through fountains, pools, streams, hydraulic organs or sounding boxes of organic materials, having the winds sound through aeolian harp-like constructions, and making birds audible, would create the right atmosphere. In 'temples of silence,' recreating the stillness of former libraries and churches, citizens would additionally be invited to contemplate and enjoy a silence that, if never absolute, could be positive.

While the idea to redesign local soundscapes has been taken up widely, especially by artists, a connection to local ordinances has not been systematically explored. I will discuss this connection as a critical next step in the suggestions recently made by urban sound designer Jordan Lacey and sound art theorist and practitioner Sabine Breitsameter. Lacey notes that the continuousness of urban noise creates a boring soundscape of "relentless similarity" (2015, n.p.), and Breitsameter claims that cities are increasingly each other's sound-alikes, because similar ways of "production, management, and consumption" lead to "analogous sound environments" (2015, p. 145). Both scholars see important roles for artists: they can reintroduce sonic diversity. Lacey suggests to not retreat into silence, but support installations that "reshape urban noise into evocative listening experiences" (2015, n.p.). Breitsameter, however, warns to not relegate listening to artistic activities only, but to educate people into "[a]esthetic listening" in everyday life (2015, p. 150). Local *ear*donances, so I would like to add, can create the legal-spatial conditions to listen aesthetically, but *also* to create spaces that leave it up to citizens themselves whether and how to prioritize listening experiences. Before I can elaborate on this, however, I first need to return to the history of anti-noise activism, and the uptake of Schafer's ideas in this world.

Sound Waves of Noise Abatement

The works resulting from the World Soundscape Project were published amidst a flurry of noise abatement campaigns and societies that had their origins in the 1950s and 1960s. In a recently issued chapter in the Routledge Companion to Sound Studies edited by Michael Bull, I have situated these societies and their campaigns in the second of three waves of twentieth-century anti-noise activism (Bijsterveld, 2018).

The first wave started a few years after the turn of the century. The New Yorker Julia Barnett Rice established the Society for the Suppression of Unnecessary Noise in 1906, philosopher Theodor Lessing raised the Deutscher Lärmschutzverband, or the German Association for Protection from Noise in 1908, and the British medical man Sir Thomas J. Horder followed suit with the Anti-Noise League in 1933 (Bijsterveld, 2008; Mansell, 2017; Thompson, 2002). Most of these societies aimed at banning needless, chaotic noise that bothered the more vulnerable groups in society. For Rice, these groups were ill people staying in hospitals as well as school children. In practice, however, her work and that by Lessing and Horder protected the peace and quiet of professional elites in particular. Neurasthenia induced by chaotic noise, so was the idea, endangered these mental workers' very existence. This was also noticeable in the types of noise the early noise abatement societies considered "unnecessary": the sounds of riverboat steam whistles, automobile horns, garbage collection, gramophone and radio were seen as health hazards, whereas industrial noise was left untouched. Its presumed rhythmical character was considered less dangerous than arrhythmical noise. In the Netherlands, the Sound Foundation, established by engineers, and involved in acoustical research as well as the practice of noise abatement, collaborated with motorist societies against traffic noise. By educating pedestrians and cyclists to use the road as predictably as possible, their campaigns resulted in a reduction of hooting. The other effect, however, was that roads were now cleared from everything that stood in the way of the automobile. Motorist traffic and transport thus smoothened instead of slowing down.

The second wave broadened the scope of what had to be protected from an ever-larger portion of society to the environment as such. While many of the early twentieth-century and interwar societies had perished, new ones were established once the most immediate effects of World War II had been dealt with. In Germany, the Deutscher Arbeitsring für Lärmbekampfung (DAL), or German Working Group for Noise Abatement was established in 1952. In the UK, the Noise Abatement Society (NAS) started its work in 1959, and the Nederlandse Stichting voor Geluidshinder (NSG), or Dutch Noise Abatement Society, was raised in 1970 (Bijsterveld, 2008, p. 184). Industrial society was no longer taken for granted, however—on the contrary. A more structural reform of capitalism was now called for by many noise abating activists. The noise of aircraft was considered as one of capitalism's excesses. Yet while noise abatement societies in the US successfully campaigned against the sonic boom of supersonic aircraft, and German anti-noise protesters violently protested against the extension of airports, flying became big business. This was accompanied by a shift in the noise abatement movement's view on urban life. In the first wave of noise abatement, the city had been presented as a place from which poor people could not escape, and from which the rich did not want to flee given their professional ambitions. The second wave, however, defined the city as a place one should leave behind in favor of rural life. Hence the introduction, by the end of the 1970s, of silence areas in The Netherlands—areas that were commonly located in the country (Bijsterveld, 2008, pp. 245–246). This expressed both the urge for environmental preservation and a more nostalgic longing for a tranquil country life.

The third wave, starting in the 1990s, radiated a more ambiguous imagination of urbanism. As living in metropoles had grown increasingly popular, the sound of the city was now both a troubling aspect and a simple fact of life. Many local noise abatement committees kept working in the tradition of the first wave of noise abatement campaigns, focusing on banning single sources of noise, ranging from the noise from leaf blowers and piped music to car alarms and aircraft. But alongside these committees, a noise awareness movement entered the scene that fostered concentrated listening to the acoustic qualities of the environment, stimulated an interventionist approach to city sound through sound design, *and* a more preservationist take that celebrated characteristic city sounds through soundwalks and soundmaps. The World Listening Day, organized annually by the World Forum for Acoustic Ecology (WFAE) since

2010, on Schafer's birthday (July 18), is part of this game. Meanwhile, the national noise abatement organizations that had been established in the 1950s and 1960s professionalized their work by rebranding themselves into service and lobby organizations that were increasingly active on the European level.

Schafer's work both reflected and contributed to the environmental take on noise pollution. He borrowed "heavily from the language of the environmental movement," so Ari Kelman (2010, p. 216) has rightly claimed. Schafer added its own twist by focusing on acoustic design, and by affording composers a brand-new role in it. Some used ear cleaning exercises to better prepare amateurs for performing avant-garde music, as the Dutch composer Tera de Marez Oyens did in *Working with Modern Sounds* (1978). Many of them created soundscapes to enrich particular cityscapes, and increasingly with ears for the potential beauty in urban sound: Lacey and Breitsameter's suggestions are in line with this trend as well. Yet what about the post-war noise abatement societies or their successors?

Sonic Awareness in Present-Day Noise Abatement Societies

To be able to find out whether and how these societies draw on Murray Schafer's legacy in more detail, I narrowed my selection down to three of the post-war Western European noise abatement societies that I had also studied in *Mechanical Sound*. These are, as announced above, the Noise Abatement Society (NAS) based in the United Kingdom, the Dutch Foundation against Noise Nuisance (NSG) and the Working Group on Noise of the German Acoustical Society (ALD).[4] I studied their websites by using the search terms Raymond Murray Schafer, soundscape, soundwalk and sound art, as well as their translations to Dutch and German.[5] In addition, I analyzed newsletters, new blogs and magazines published by these societies in the years 2011 through 2015.[6]

The glossy NAS Ezine *SoundScape* immediately reveals that Raymond Murray Schafer's work is known among the society's leadership. In fact, Schafer's writings are said to have directly inspired the magazine's title (2011, p. 26, 2012, p. 8). His message is visible in the magazine's motto as well: "SoundScape: where life sounds good." Indeed, Schafer's work is cited (2011, p. 43, 2012, p. 10, p. 69), his publications celebrated as defining moments in the history of noise abatement (2012, p. 29), and his key notions are both explained (2012, p. 8) and used, for instance, in an overview of the "soundmarks" of the Olympic Games in London, with the three-minute ringing of bells across the entire country as a highly welcome example (2012, pp. 10–12, pp. 31–39).

It does not mean that the NAS does not cling to any of the more conventional strategies, such as collecting complaints and offering free support and advice, be it through new formats. As of 2006, for instance, it has a National Noise Helpline. The helpline's statistics show that "neighbor noise" has been the largest category of concerns, with poor insulation of residential housing as one of the most significant causes (NAS, 2011, pp. 6–8). It also refers to traditional campaigns such as those against the third Heathrow runway (2011, pp. 44–45), a quiet nighttime delivery program (2011, pp. 46–49), a sound dining campaign (2011, pp. 50–52) and a Noise Manifesto for a quiet revolution in building (NAS, 2013).

But most of the campaigns clearly bear a Schaferian touch. This does not only hold for a series of 'Noise Oscars,' ranging from prizes named after John Connell—the business man who established the NAS—such as the John Connell Local Authority, Innovation and Technology Awards, Quiet Approach™ Award and the European Soundscape Award (NAS, 2011, pp. 18–19, 2012, pp. 65–67; NAS News, 2011). It is also evident in campaigns such as 'Love Your Ears' aimed at making youngsters aware of the dangers of extensive and loud personal stereo use (2011, pp. 16–17, 2012, pp. 6–7), or articles on the importance of schools with properly designed classrooms, such as classrooms with lower reverberation times (2011, pp. 10–15).

In line with its founder's background, the NAS doesn't shy away from using its links with industry in cultivating this positive approach. The Quiet Mark, for instance, is another award-like invention celebrating products that contribute to noise control. In addition, the magazine is full of advertisements for such quiet products, and features articles and interviews that are close to being advertorials, such as those written by staff employed by Rockwool on acoustic insulation materials for renovating a flat (NAS, 2011, pp. 22–25) or Brigade Electronics that sells white sound reversing alarms that are claimed to be less annoying than conventional alarms (2011, pp. 54–55).

Even more important, however, is the way in which Schafer's focus on acoustic design is noticeable in many of the Noise Abatement Society's campaigns. The *Soundscape* journal's second issue aims to create a "world at peace," and "to craft positive soundscapes" (2011, p. 3). Interestingly, the city is acknowledged as the place in which the urban populations' search for relief can be answered by proper acoustic design. Schafer's notions are not always cited uncritically, however. One article, for instance, presents the thoughts and work by sound artist Emma Quayle. She brings Schafer's opposition between the hi-fi aural environment of rural soundscapes and the lo-fi environment of the city into mind, but notes that a hi-fi environment can be disruptive, for instance, when the sound of a wind turbine is audible on the countryside. She therefore suggests to replace it by the opposition between human and non-human sounds, and to seek a balance between the two. Her own sound installation, *Organ of Corti*, has been awarded with a John Connell Innovation Award, and is an instrument with transparent tubes that "'recycles noise from the environment," framing "the sounds already present … in a new way" and also "challenges expectations of what might constitute a piece of music" (NAS, 2011, p. 21).

It expresses the open ears, tolerant take on urban sound that is typical of the 1990s' twist in the uptake of Schafer's work. This is also visible in articles about Tuned City projects, such as the one in Tallinn which celebrated Estonia's rich sonic history, presented installations that "transformed urban noise into tuned sounds," and had participants discover the quiet, "usually unheard" sounds of the city (2011, pp. 26–27). Another example is that of Tuned City Berlin, presenting Berlin as a city with a relatively quiet soundscape given its preservation of trees, bushes and city wildlife, and debating the acoustic futures of spaces such as Tempelhof (NAS, 2012, pp. 20–24). Also typical is an item about the project Sounding Brighton that explored "new ways of listening to local soundscapes and new ways of tackling noise and improving local soundscape quality" through commissioned sonic artworks (NAS, 2011, p. 42; NAS News, 2012). One of those was a 3D soundscape audible amidst two rows of speakers that had been situated "in the heart of Brighton's night life," with compositions of

> both recorded and live sounds from a kiosk at the side of the street. The soundscape presented a contrast to the raucous disharmony so frequently heard in lively areas at night, and connected with visitors to the area and residents, as well as those exiting in clubs. (2011, p. 42)

Even though the artwork drew on technologies that were not exactly Schaferian, the harmony and communal connections it was thought to evoke are fully in line with his ideas. And the Peace Gardens in Sheffield City, awarded with a John Connell Award, are applauded for the "acoustically welcoming and pleasant" tones of their fountains and water cascades—constructions referring to the significance of water in Sheffield's local history (NAS, 2012, p. 59).

The German Arbeitsring Lärm der Deutschen Gesellschaft für Akustik (ALD), or Working Group on Noise of the German Acoustical Society (ALD), established in 2009, is the official successor of the Deutscher Arbeitsring für Lärmbekämpfung (DAL), or the German Working Group for Noise Abatement (DAL) that has its roots in the 1950s. The DAL had, and the ALD still has, strong links with the sciences. The DAL was established by physicians and lawyers, and supported by the Verein Deutscher Ingenieure, or the Association of German Engineers, while the ALD is embedded in the world of German acoustics.

These backgrounds are highly visible in the way the ALD operates today. It says to draw on professional expertise and independence to improve the protection against noise in both Germany and Europe at large, and to enhance noise awareness. The ALD Newsletters are full of items that critically discuss noise caused by cars, railway transport, aircraft, wind turbines, heat pumps or other sources in terms of their measured sound level emission, state and federal environmental legislation, and news on research. Typical for the German situation are the fierce debates of noise caused by children playing near daycare venues or in gardens. In 2011, the German government tried to end this debate by exempting children from noise regulation, defending the principle that children have the right to make noise as a phase in their development (ALD, 2011, p. 6). The ALD did not give up on the issue, however, assisted by professor of psychoacoustics Brigitte Schulte-Fortkamp who has introduced the notion of soundscaping to take the subjectivities of perception into account, engage citizens in discussions and mediate between them if necessary. Her scholarly and societal work was awarded with the European Soundscape Award 2012, issued by the European Environment Agency, the NAS and the NSG (ALD, 2011, pp. 20–21).

If this is one way in which the legacy of acoustic ecology can be traced, the other is the transformation of an ALD-instrument, the *Lärmspaziergänge*, or noise walks, into soundwalks. These noise walks were originally meant to invite local authorities, residents and local experts to measure noise, and to enhance noise awareness through information-rich talks before taking off for the actual walk in a particular area. Since at least 2013, however, the ALD has added elements of the "soundwalk" by also visiting spots that are locally known for their tranquility, and by combining objective measurements with subjective ratings by those participating in the event. Examples of the relatively still places visited are the library of Darmstadt, a park in Berlin-Pankow and a courtyard in Berlin Kreuzberg (ALD, 2013, pp. 13–15, 2014, pp. 52–53). This courtyard's sonic atmosphere has been qualified as "peaceful," "nearly monastic" and "idyllic."[7] Such walks have usually been organized in the context of the International Noise Awareness Days linked to the Center for Hearing and Communication (CLC) rather than the World Listening Days initiated by the WFAE. The effect of Schafer's acoustic ecology approach is noticeable nonetheless.

In contrast, the work done by the Dutch Society Against Noise Nuisance seems to be nearly untouched by the legacy of acoustic ecology, at least at first sight. Its NSG newsletters remain full of traditional noise abatement topics such as neighbor noise, road traffic noise, amplified music in pubs and clubs, and the hard-to-grapple issue of low frequency noise. Just like the NAS, the NSG also works with awards, such as the "Golden Decibel Award" for initiatives that lead to increasing noise control (NSG, 2012, p. 6). In addition, it promotes the use of relatively silent E-scooters among youth, and a proper choice of car tires by motorists. Perhaps this fairly conventional approach is due to the fact that the foundation considers itself as a first aid for noise issues. The only examples that speak against the foundation's classical noise abatement approach are brief references to the European Soundscape Awards, a visit to a work of sound art in the context of a meeting on sound policies at the municipal level, and the organization of a "Masterclass Soundscape," including a "soundwalk," in the context of a draft Dutch law on the environment, the *Omgevingswet* (NSG, 2013, p. 4).[8]

This masterclass, however, responded to the fact that Dutch politics itself had taken the notion of soundscape on board. In fact, it seems to follow Schafer's intentions quite closely, as it suggests to move away from framing noise abatement politics merely in terms of measured levels of sound, but focus instead on forms of acoustic design ("soundscaping") that take the perception of (un)pleasant sounds by local residents into account, while allowing for higher maximum sound levels.[9] This also prompts a reflection about my own approach in this chapter—as it shows that noise abatement is not restricted to the campaigns and events organized by noise abatement societies. So, in some cases, Schafer's work found its way into legislation in spite of noise abatement societies' neglect of his approach.

Local *Ear*donances and Schaferian Solutions

What has not happened, but would be worthwhile to explore, is to return to Schafer's initial interest in local bylaws or ordinances, and turn them into timely *ear*donances. It would be a Schaferian solution to not only redesign urban surroundings in such a way that soniferous gardens would offer sonic relief but also to rewrite local ordinances into *ear*donances that make residents entitled to spaces allowing for undisturbed listening or even non-listening time. This is not to say that a reduction of noise as its source is not the best solution—it is still highly practical—but as long as society keeps adding sounding commodities and festivals to everyday life, other approaches might be helpful as well. Local *ear*donances would, in my view, be especially fruitful for finding a way out of the tensions that might explain the growing use of Schafer's work from the 1990s and after: citizens' increasing preference, for various reasons, to live in cities (and enjoy its sounding offerings) and an increasing noise awareness. Noise abatement societies' inclusion of writings about pleasant soundscapes and positive strategies, even in various degrees, can be interpreted as a response to that tension. If they would simply bash cities, they might lose contact with large portions of society, especially the younger groups.

But how would this translate in local *ear*donances? Let me start with two examples. Both Amsterdam and Berlin are struggling with an increase of urban music festivals. Whereas music festivals attract tourists and young professionals, many city dwellers complain. With twenty festival hotspots, the Amsterdam city council wants to accomplish a more balanced distribution of music festivals across the city to reduce the noise nuisance in the most heavily affected residential areas. The festivals are also required to undertake measurements on the façades of residential housing to ensure that they stay below particular maximum levels and use the best noise control technologies available. These are, for instance, the use of containers as noise screens on festival grounds, or special sub woofers that deaden the bass at the backside of the stage (Zwetsloot, 2017). Changing locations, however, may very well create new insecurities for citizens and may not necessarily lead to less nuisance and complaints; the other interventions are likely to be more effective, even if quiet nights are out of reach for festival neighbors.

Berlin came up with another remarkable strategy. When the September 2016 edition of the traveling music festival Lollapalooza was upcoming, it had to shift its location from the former airport Tempelhof to Treptower Park. This was because Tempelhof had become the temporary residency of thousands of refugees. Treptower Park residents protested fiercely in two citizens' meetings against the expected noise nuisance in the festival weekend, given the 45 bands, 5 stages and 70,000 expected visitors, even though the festival had to stop at 10 PM on one day and 11 PM at the other (Goldstein, 2016, p. 10). Their protests were joined by those ambassadors of former Soviet states who deplored the raucous disturbance of the peace at the Soviet monument, around which thousands of Russian soldiers had been buried in the wake of World War II. The protests did not result in banning the festival, according to one newspaper because the "senate ... rejoices in the image of Berlin as world metropole." Yet Lollapalooza had to offer alternative shelters, that is, hotels, to over 1600 residents who had to cope with a sound load higher than 80 decibel (Strauss, 2016, p. 12).

Wouldn't it be better, rather than forcing citizens to go to shelters at the last minute, as if festivals are natural disasters or wars, to create structural islands of relief in which citizens can stay for free or for very low fees, and at their own initiative, as soon as they suffer from festivals or other events? What about local ordinances that give residents the right to book a few nights in soniferous gardens with circular sound-insulated hotels around them? If cities are considered attractive because of their festivals, but noise awareness reduces the tolerance of those inhabiting them, what about extending leisure activities to the right to stay in secluded serenity gardens? In parks that would be free from festivals and would have smartly designed buildings that offer peace and tranquility? And that would, at the very same time, be experimental spaces—spaces in which architects and industrial designers would try out new materials? Local *ear*donances, in my argument, would then be regulations creating time and space for undisturbed listening, or even for experiences that do not require conscious listening at all. After all, educating citizens into aesthetic listening, or presenting them with installations that aestheticize noise—in whose ears?— is not without paternalism, and are attempts to intervene at the deepest individual level. Collective initiatives might actually leave citizens with more hearing and breathing space. And wouldn't local authorities be in for it? Creating cooperatives designing and building serenity spaces might speak as much to youngsters as music festivals.

Conclusions

That Raymond Murray Schafer did not add a model city bylaw to *The Tuning of the World*, even though he would have been able to do so, was entirely in line with his intention to focus on environmental acoustic design rather than noise abatement. And as most local ordinances sent in as response to the worldwide urban questionnaire centered on banning particular sources of sound, they did not suit his main purpose either. But the latest wave of noise awareness initiatives, the immense popularity of living in urban areas, and the societal tensions concerning the sound levels generated by amplified music within the confines of the city, give reason to revisit the local ordinances, and to create new, more positively formulated ones.

Such local *ear*donances should ensure some easily accessible and smartly designed spaces for acoustic retreat and enrichment for those whose hearing selves are not able to cope with the noise of today's urban celebrations. Again, this is not to say that I consider a reduction of noise at its source, as acousticians would formulate it, useless. I also see the dangers of the new Dutch approach, in which maximum allowed sound levels are raised with the argument that sound design can take the acoustic identity of urban areas into account, especially as long as consultants do not know what this means. In the current situation, however, the well-to-do can flee their urban houses if the festival seasons get too intense, while the less well-off are dependent on last minutes interventions by judges or just have to live with the noise. Local *ear*donances that entitle them to make use and temporarily stay at especially designed and dedicated places of stillness would not only help these urban residents, but also make acoustic ecology actually work in a world of metropoles.

Notes

1. Schafer and his WSP colleagues also publish *A Survey of Community Noise By-laws in Canada* in 1972 (see The World Soundscape Project, http://www.sfu.ca/~truax/wsp.html, last accessed October 6, 2017).
2. WSP Archives, Vancouver, File Cabinet 3, Drawer 1, International Bylaws, Letter August 23, 1973, R. Murray Schafer to "Your Worship," pp. 1–2.
3. Examples are the letters sent in by Leeds (UK), Essen (Germany) and Bern (Switzerland). WSP Archives, Vancouver, File Cabinet 3, Drawer 1, International Bylaws, Letter City of Leeds Public Health Department to

Professor Schafer, March 4, 1974; Letter "Stadt Essen" to Simon Fraser University, January 9, 1974; Letter Dr. G. Schmezer, City Delegate for Information Bern, August 28, 1973.

4. For the NAS, see http://noiseabatementsociety.com/; for the NSG, see http://www.nsg.nl/nl/home.html; and for the ALD, see http://www.ald-laerm.de/ald/wir-ueber-uns/ (Last accessed at September 22, 2017).
5. In German, these search terms are *Klanglandschaft, Lärmspaziergang* and *Klangkunst*. In Dutch, the suitable words are *geluidslandschap, geluidswandeling* en *geluidskunst*.
6. With regards to the Noise Abatement Society, I studied the news blogs published on the NAS website between 2011 and 2015, as well as the Ezine *SoundScape* (the issues 02 and 03 from 2011 and 2012 respectively), available at http://noiseabatementsociety.com/soundscapeezine/past/ (Last accessed at September 22, 2017). Concerning the NSG, I analyzed their newsletters for donators (*Donateursblad*) published between 2011 and 2015, available at http://www.nsg.nl/nl/nsg-donateursblad.html (Last accessed at September 22, 2017).

   As to the ALD, I examined the ALD newsletters published between 2011 and 2015. These newsletters are available at http://www.ald-laerm.de/publikationen/newsletter-des-ald/newsletter-archiv/ (Last accessed at September 22, 2017).
7. Kay S. Voigt, Bericht zum ALD-Lärmspaziergang/Soundwalk http://www.ald-laerm.de/fileadmin/ald-laerm.de/Projekte_des_ALD/2013/ALD-Laermspaziergaenge/Bericht_Soundwalk_Kreuzberg_Endversion.pdf, at p. 13 and p. 18 (Last accessed October 26, 2017).
8. See http://www.nsg.nl/nl/masterclass_soundscape.html (Last accessed October 26, 2017).
9. See https://www.mp.nl/actueel/soundscaping-als-instrument-voor-een-aangenaam-leefklimaat (Last accessed October 26, 2017).

References

ALD. (2011, April 21). *Newsletter Nr. 7*. Retrieved from http://www.ald-laerm.de/publikationen/newsletter-des-ald/newsletter-archiv/

ALD. (2013, September 06). *Newsletter Nr. 13*. Retrieved from http://www.ald-laerm.de/publikationen/newsletter-des-ald/newsletter-archiv/

ALD. (2014, July 04). *Newsletter Nr. 14*. Retrieved from http://www.ald-laerm.de/publikationen/newsletter-des-ald/newsletter-archiv/

Bijsterveld, K. (2003). 'The City of Din': Decibels, Noise and Neighbors in the Netherlands, 1910–1980. *Osiris, 18*, 173–193.

Bijsterveld, K. (2008). *Mechanical Sound: Technology, Culture, and Public Problems of Noise in the Twentieth Century*. Cambridge, MA: The MIT Press.

Bijsterveld, K. (2018, forthcoming). Sound Waves of Protest: Noise Abatement Movements. In M. Bull (Ed.), *Routledge Companion to Sound Studies*. New York: Routledge.

Breitsameter, S. (2015). Ordering of Sounds: The Homogenization of Listening in the Age of Globalized Soundscapes. In C. Weaver, D. Klassen, & J. Klassen (Eds.), *Sound in the Land: Music and the Environment*, Special Issue, *The Conrad Grebel Review, 33*(2), 142–150.

Goldstein, P. (2016, August 30). Gartenarbeit für den Rock 'n' Roll. *Berliner Morgenpost*, 10.

Kelman, A. Y. (2010). Rethinking the Soundscape: A Critical Genealogy of a Key Term in Sound Studies. *The Senses and Society, 5*(2), 212–234.

Lacey, J. (2015, August 19). We Need a New Relationship with Urban Noise. *The Conversation*. Retrieved from https://theconversation.com/we-need-a-new-relationship-with-urban-noise-46207. Last Accessed March 29, 2018.

Mansell, J. G. (2017). *The Age of Noise in Britain: Hearing Modernity*. Urbana, IL: University of Illinois Press.

Marez Oyens, T. (1978). *Werken met moderne klanken: Samenspel voor stemmen en instrumenten*. Haarlem, Netherlands: De Toorts.

NAS. (2011). *SoundScape – Issue Two*. Retrieved from http://noiseabatementsociety.com/soundscapeezine/past/

NAS. (2012). *SoundScape – Issue Three*. Retrieved from http://noiseabatementsociety.com/soundscapeezine/past/

NAS. (2013, February 19). *Noise Manifesto Launched by Noise Abatement Society and Rockwool Calling for Urgent Planning Reform to Start 'Quiet Revolution' in Building Design* [Blog Post]. Retrieved from http://noiseabatementsociety.com/2013/02/19/noise-manifesto-launched-by-noise-abatement-society-and-rockwool-calling-for-urgent-planning-reform-to-start-quiet-revolution-in-building-design/

NAS News. (2011, November 09). *Local Authorities Sweep the Board at Awards* [Blog Post]. Retrieved from http://noiseabatementsociety.com/2011/11/09/local-authorities-sweep-the-board-at-awards/

NAS News. (2012, June 20). *Brighton & Hove and NAS Are Sounding Brighton – "Exploring Practical Approaches Towards Better Soundscapes"*. Retrieved from http://noiseabatementsociety.com/2012/06/20/brighton-hove-and-nas-are-sounding-brighton-exploring-practical-approaches-towards-better-soundscapes-4/

NSG. (2012, July). *Jaargang* [donateursblad]. Retrieved from https://www.nsg.nl/nl/nsg-donateursblad.html

NSG. (2013, July). *Jaargang* [donateursblad]. Retrieved from https://www.nsg.nl/nl/nsg-donateursblad.html

Radovac, L. (2012). The "War on Noise": Sound and Space in La Guardia's New York. In K. Keeling & J. Kun (Eds.), *Sound Clash: Listening to American Studies* (pp. 289–316). Baltimore: The Johns Hopkins University Press.

Schafer, R. M. (1973). *The Book of Noise*. Wellington, New Zealand: Price Milburn.

Schafer, R. M. (Ed.). (1977a). *European Sound Diary*. Burnaby, Canada: Simon Fraser University.

Schafer, R. M. (1977b). *Five Village Soundscapes*. Burnaby, Canada: Simon Fraser University.

Schafer, R. M. (1977c). *The Tuning of the World*. New York, NY: Knopf. Reprinted as *The Soundscape: Our Sonic Environment and the Tuning of the World*. Rochester, VT: Destiny Books, 1994.

Sterne, J. (2013). Soundscape, Landscape, Escape. In K. Bijsterveld (Ed.), *Soundscapes of the Urban Past: Staged Sound as Mediated Cultural Heritage* (pp. 181–193). Bielefeld, Germany: Transcript Verlag.

Strauss, S. (2016, September 8). Lollapalooza darf loslegen. *Berliner Zeitung*, No. 211, 12.

Thompson, E. (2002). *The Soundscape of Modernity. Architectural Acoustics 1900–1933*. Cambridge, MA: MIT Press.

Zwetsloot, J. (2017, July 10). Herriestoppers. *De Volkskrant*, (Economie), 23.

CHAPTER 5

Multimodal Scholarship in World Soundscape Project Composition: Toward a Different Media-Theoretical Legacy (Or: The WSP as OG DH)

Jonathan Sterne

In this chapter, I reconsider the World Soundscape Project's (WSP) media practices as media theories in action (or The WSP as OG DH).[1] I listen to and analyze two canonical works, Hildegard Westerkamp's *Kits Beach Soundwalk* (1989) and Barry Truax's *Riverrun* (1986), to consider them as examples of media theory in the sonic register. I argue, anachronistically, that these works are examples of *multimodal scholarship* before that term came into existence. In doing this, I hope to help bring sonic histories and practices more fully into discussions of the digital humanities while also offering an alternative to the ways in which media theory coming out of the World Soundscape Project is usually discussed.

As Tara McPherson explains, multimodal scholars bring together "databases, scholarly tools, networked writing, and peer-to-peer commentary

J. Sterne (✉)
McGill University, Montreal, QC, Canada
e-mail: jonathan.sterne@mcgill.ca

M. Droumeva, R. Jordan (eds.), *Sound, Media, Ecology*,
Palgrave Studies in Audio-Visual Culture,
https://doi.org/10.1007/978-3-030-16569-7_5

while also leveraging visual and aural media that so dominate contemporary life" (2009, p. 120). They consider how arguments are experienced or felt in more "sensory-rich" spaces than writing alone. While traditional scholarship privileges writing as its ideal mode of dissemination, multimodal scholarship might juxtapose sounds and images with writing; it might produce nonlinear forms of writing aided by digital navigation; or it might use media like sound recordings, videos, or games, where writing takes on a secondary or supporting role. In her discussion of multimodality, Kathleen Hayles refers to the "full range of visual images, graphics, animations and other digital effects," arguing that the best projects "have emotional force as well as conceptual coherence" (2012, p. 40). Writing of digital technologies, McPherson argues that:

> hands-on engagement with digital forms reorients the scholarly imagination, not because the tools are cool or new (even if they are) or because the audience for our work might be expanded (even if it is), but because scholars come to realize that they understand their arguments and their objects of study differently, even better, when they approach them through multiple modalities and emergent and interconnected forms of literacy. The ability to deploy new experiential, emotional, and even tactile aspects of argument and expression can open up fresh avenues of inquiry and research. (2009, p. 121)

Yet the theories and tools of the digital humanities have been heavily visualist and textualist in orientation, in part because of the digital humanities' institutional histories, and in the operational privilege of text over sound in many digital media applications (Lingold, Mueller, & Trettien, 2018, pp. 9–10). Even massive audio archives and collections require metadata in the form of text to make sounds searchable, identifiable, and retrievable (Morris, 2015). While multimodal publication tools like Scalar and Manifold have audiovisual capabilities, their video players have been developed long before dedicated audio players, which reflects the disciplinary interests of the scholars building and using them. In part, this bias reflects the wider commercial priorities of web application developers, who also often treat audio as a secondary concern.

If the digital humanities is to be understood as emanating from fields like literature, film studies, and visual culture, there is a story to be told about overcoming the opposition between criticism and production because of those fields' historical emphasis on that opposition. But not all disciplines work that way. In fields like engineering and computer science, that opposition makes no sense. So too for acoustic ecology, which was heavily influenced by the ideas behind modernist composition practices. To foster new orientations to music and sound, intellectually oriented composers appropriated intellectual trends from outside music: from information theory (Karlheinz Stockhausen, Pierre Schaeffer, Iannis Xenakis), to lesbian-feminist utopianism (Pauline Oliveros), to ecology (Hildegard Westerkamp, R. Murray Schafer), to Western digestions of Buddhist and Taoist philosophy (John Cage).

The compositional, collecting, and radio practices of the World Soundscape Project thus fit into a longer tradition of technologized humanism. As both McPherson (2018) and Hayles suggest (and they are far from alone in this respect), the move from critique to critical production is occasioned by the proliferation of digital technologies. But the humanities have long been technological. Digital tools are simply more apparent as technologies because of their relative novelty. Crotchety colleagues have sometimes responded to the millenarian rhetoric of digital humanities by positing "analog humanities" as an alternative, even borrowing from vinyl record collectors to call for a revival. I wish to imply neither nostalgia nor revivalism. In an essay on the history of humanistic uses of examples, I used the term "analog humanities" to refer to humanists' uses of analog media technologies—and the analog components of digital technologies—in academic settings and in print. The parallel with *digital humanities* is intentional; both terms refer to complexes of technologies and engagements, without specifying any particular discipline (Sterne, 2015).

As Lisa Gitelman writes, "Media history offers access to the epistemologies and interpretive practices of the humanities at a vernacular as well as scholarly or academic level [...] media aren't instruments of scholarship in the humanities; they are the instruments of humanism at large ..." (2006, p. 153). If we recognize the long-term multimodality and mediality of humanistic scholarship, then it is perhaps not radical at all to incorporate new technological modes into our work. But just as the institutional shape of education is changing—due to a mix of changes in hardware and a series of political economic shifts—so too are our orientations to our tools and, through our tools, the texts and artifacts we study and the ideas we advance.

This is where a rereading of the World Soundscape Project's media work may be useful. Unlike humanists' use of dual slide projectors for art history lectures and mimeograph machines and cameras and darkrooms for their documentary practices, the use of sound recording in academic settings was flagged as "technological" at the time because it was specialized and expensive. It is also easy to recognize as technological today because of the ways in which sound recording is *still* considered to be a specialized, technologized practice even though most people in industrial societies do it as a matter of course. The overlap between scholarship and consumerism is also important here: today, both teachers and students walk around with devices in their pockets more powerful than the entire studio at Simon Fraser University in the 1980s. Yet as I discuss in the following paragraphs, there is a powerful ideology that still treats sound recording (and audio practice) as a specialized skill, as opposed to photography or video editing.

The World Soundscape Project's media practice has *already fulfilled* some of the ambitions for the digital humanities; it could be a useful touchstone for scholars who want to make arguments through and in sonic registers. This is why, tongue in cheek, I call the WSP "O.G. DH."[2] Most often, when the WSP comes up in media theory, R. Murray Schafer's writings are held up as the group's main theoretical contribution. And in recent years, Schafer's ideas have come under heavy criticism from media theorists. His concept of schizophonia mobilizes disability stigma to argue that the mediation of sound is inherently psychologically damaging, a position that trivializes mediation and, importantly, ignores that all sounds are "separated from their sources" (Stanyek & Piekut, 2010). His concepts of noise are inherently anti-urban and work with, rather than against, notions of noise that are used to control and stigmatize racialized and minoritized populations (Blake, 2010; Radovac, 2015). His concept of soundscape is too broad, susceptible to the many critiques of landscape as a totalizing phenomenon in art history and geography (Akiyama, 2010; Thompson, 2002).

Yet a focus on Schafer's *writings* has obscured a focus on the WSP's *practices*. Considered in their fullness, these practices often directly contradict the media theories advanced in Schafer's writings. Several strands of media studies have treated media practices and technologies as operative theories: writers like Gilles Deleuze (2001), Friedrich Kittler (1999), and Wolfgang Ernst (2016) have considered time-based technologies like cinema and sound recording as philosophies that comment on the nature of temporality and subjectivity or, rather, the very possibilities of those phenomena. Lisa Gitelman (2006) has treated the sound recording as a meditation on the nature of all sorts of records and documents. Wendy Chun (2011), Alex Galloway (2012), and Sumanth Gopinath (2013) have considered software processes as theories of ideology and subjectivity. Media art histories have considered sonic artworks as philosophical meditations on the nature of capital, subjectivity, and nature (Curran, 2010–2018; Dyson, 2009; Eidsheim, 2015; Kahn, 2013; Kelly, 2009; Knouf, 2016). And my own monographs (Sterne, 2003, 2012) consider the theories of subjects that are directly built or written into auditory technologies from vibrating membranes in the nineteenth century to audio codecs in the twenty-first century.

"Theory" here refers to two different referents, which should be clear from context but are not the same thing. *Media theory* denotes an interdisciplinary field of discussion about how media work, why they matter, and so forth, just as literary theory does the same for literature and anthropological theory does the same for culture. However, a *media theory* is not simply a comment on that intellectual field. Rather, a *theory* is a description of the world with explanatory or predictive power. To confuse the former with the latter is to obscure what makes actual theories useful: they help us explain, describe, or predict aspects of the world. In the context of scholarship, "takes" on the works of other authors are commentaries, but lack that explanatory power. My commentaries on other authors in this chapter do not constitute *a theory* but may be part of the dialogue within the field of *media theory*. Conversely, I am arguing that the two pieces I examine in detail later should be understood as themselves offering theories of media and sound because they have explanatory power.

To analyze the WSP's media practice as a working set of media theories, we will have to move beyond the ways their work is usually discussed—as art and as documentary practice. The many sonic practices undertaken by the World Soundscape Project were not generally narrated as a kind of humanistic work. The three main figures I consider in this piece—Hildegard Westerkamp, Barry Truax, and, parenthetically, R. Murray Schafer—all refer to their sound work as *composition* and have labored, with great success, to have their output understood within the global economies of "serious" composed music and sonic artwork. While I have no interest in challenging the terms on which they are evaluated in that sphere, I bring a different hermeneutic to their work. In this essay, I will evaluate two exemplary works as argumentative, demonstrative, and didactic—and not as art. I will not consider their work in terms of artistic vision, originality, or integrity according to the avant-garde compositional ideals operating in their time or ours. Rather, I want to consider this work as a precursor of the kind of multimodal scholarship that is now being done under the flag of digital humanities.

Thus, this chapter considers the WSP's media work (alongside their writings, but not necessarily interpreted through them) as theoretical and pedagogical in and of itself. To do this, I offer readings that may contradict published writings and statements by WSP principals in order to reread the past to produce a story that may be more useful for the present (White, 1966). My reading is intentionally anachronistic and anti-intentional: it considers work in the 1980s through the debates of the 2000s and 2010s; it somewhat discounts the ways in which Westerkamp and Truax talked about their work in order to cast it in terms contemporary to us now, rather than to their context then. I have intentionally not asked either of them if they would imagine their older work in terms of the digital humanities. In his critique of historicism, Jonathan Rée wrote that "There is no good reason why the thought of a period should not be judged by the standards of another" (1991, p. 979; see also Leslie, 1970). That is what I aim to do here, in the service of using past work as a resource for imagining possible futures.

Hearing and Playing *Kits Beach Soundwalk* Didactically

Hildegard Westerkamp's *Kits Beach Soundwalk* is one of her most famous works, showing up in reviews of avant-garde music in the twentieth century, in catalogs of WSP practice, and on course syllabi.[3] One notable reading of the piece comes from Tim Rutherford-Johnson, in *Music After the Fall,* a review of avant-garde composition after 1989. Rutherford-Johnson finds it exemplary because it does not rely on a score, yet it reflects a "painstaking and reflexive period of composition,"[4] evidenced by careful placements of sounds so they do not mask one another (p. 4).

To understand my reading of *Kits Beach* as an example of *didactic* work, it is helpful to understand how it built on Westerkamp's radio program *Soundwalking*, which was featured on Vancouver's co-operative radio station. Westerkamp wrote that she wanted to create radio that "inspires us to invent; […] refreshes our acoustic sensitivity; […] stimulates listening; does not repeat; […] encourages us to sign or to speak, to make radio ourselves […] instead of merely broadcasting at us" (Westerkamp, 1994, p. 89). In some ways, this is a fairly standard critique of the unidirectional format of late twentieth-century commercial broadcasting, but unlike many of the mass culture critics who had been making the same point about overcoming passivity for some time (most famously Adorno and his critique of the listener), Westerkamp framed her radio work as pedagogy that promoted an alternative social vision of sound, and of media.

> I was attempting to make radio a place of environmental listening by broadcasting the soundscapes that listeners experienced in their daily lives. With that I had hoped to create a state of resonance within listeners so that when they encountered sounds in the actual environment, recollations of the radio broadcast would alert them to the soundscape in which they lived—creating participating listeners, that is, listeners of the broadcast who could then also be receptive to the soundscape as a whole. (1994, p. 88)

Westerkamp clearly describes her radio work as related to music composition and aesthetic practice, yet one is struck by the similarity between the goals in Westerkamp's broadcasts and the kinds of things teachers like me hope to accomplish in undergraduate courses on sound. It is ultimately a pedagogical and orientational goal: for the audience to be more aware of its environment; for listeners, through instruction and practice, to develop the skills and techniques in order to listen to it; and for listeners to be motivated to engage with their environments in ways other than those prescribed by commercial media or common sense. One can even read the program like a syllabus, where the problem of listening to the environment was explored from a different location each week. For different episodes, Westerkamp and her tape recorder and microphone visited "a shopping mall, park, zoo, factory, residential area located under a flight-path and the streets of Vancouver" in order to "present everyday acoustic environments from a new angle" (Westerkamp, 1994, pp. 89–90). In the composer's note for *Kits Beach*, after summarizing her *Soundwalking* radio work, she writes that "*Kits Beach Soundwalk* is a compositional extension of this original idea" (Westerkamp, 1989) in that it combines environmental recordings, studio production techniques, and Westerkamp's didactic narration.

Readings of the piece by Brandon Labelle and Tim Rutherford-Johnson focus on a pivotal moment (for the fullest textual exposition of the piece, see McCartney, 1999, pp. 218–222). After setting the scene of a calm winter morning in January at Kits Beach in Vancouver, she focuses on the sounds of the barnacles: "I'm trying to listen to those tiny sounds in more detail now. Suddenly the background sound of the city seems louder again.… Luckily, we have bandpass filters and equalizers. We can just go into the studio and get rid of the city, pretend it's not there. Pretend we are somewhere far away." Rutherford-Johnson describes this moment of the piece as one where it "does exactly that: the city's low road is filtered away, leaving only the click and suck of the barnacles" (p. 14). The piece is meant to draw listeners' attention to those particular sounds, but it equally can draw attention to Westerkamp's own agency as an engineer in the studio. This is a crucial point because the combination of commentary and signal processing draws attention to the artifice of the recording itself. Brandon Labelle notes something similar when Westerkamp says, "I could shock you or fool you by saying the soundscape is this loud," as she turns up the volume of the background sounds, "but it is more like this," as she drops them down again. Labelle considers this a double move: "Such play opens up a space within the recording that accentuates her actual presence in the real environment while revealing the compositional components of constructing what we are hearing" (Labelle, 2015, p. 206).

Let's extend Labelle's observation. The performative doubling at these moments in *Kits Beach Soundwalk*—Westerkamp was actually there and Westerkamp actually manipulated the recordings—illustrates the fundamental phenomenological contradiction of recordings: they can be experienced as *both* indexical signs and semantico-referential (which is to say arbitrary) signs (Peirce, 1955). Understood socially and mechanically, audio recording and reproduction systems are lengthy chains of cause and effects. From the sound of Westerkamp's voice, a listener infers that she once spoke. From the sound of the barnacles, a listener could infer that she was once near them, even though neither is *necessarily* the case. This is the case for recording as indexical. Westerkamp's vocal and gestural performance mounts a case for the arbitrary nature of recording: the relation between foreground and background is entirely malleable and its meaning is further inflected by commentary. In *Kits Beach Soundwalk,* that commentary exists within the recording; but meaning for sound recordings can just as easily emerge paratextually, which is to say outside the recording itself, as in the case of music genres (Brackett, 2016) or of the silence in the Nixon tapes (Kim-Cohen, 2018). All of these propositions are manifestly performed and narrated in *Kits Beach.* They are expressed didactically, which is to say the piece can be heard to be making an argument.

Compare my explication of *Kits Beach*, as a kind of media theory, with Jason Stanyek and Ben Piekut's (2010, all quotes that follow are from p. 19) written theory of recording as media form—not to authorize her text (it doesn't need authorization) but rather to think across two modalities. They draw attention to the artifice of recording through a suite of characteristics that they can name and describe but not perform in a print medium: "*Revertibility* refers to a temporal process of undoing a work of recording in some way, whereby presumptive wholes can be disarticulated and taken back to a prior stage in a process of assemblage, upsetting straightforward, cumulative forms of co-labor." When Westerkamp plays with filtering, she demonstrates *and explains this process* in her own terms. When Westerkamp calls attention to her editing and microphone placement, she performs a kind of *recombinatoriality,* which Stanyek and Piekut describe as "the capacity toward articulating what are taken to be discrete, non-identical parts into new arrangements." Another point of overlap is in Stanyek and Piekut's use of the term *deadness* as an alternative to "the unhelpful and overvalued schism between presence and absence that undergirds" many media theories (p. 20). Westerkamp's didactic shuttling between indexical and arbitrary modes of signification makes the same point.

This is echoed in some of the listener responses that Andra McCartney cataloged, in what is no doubt the most extensive research on the piece to date. McCartney played the piece for Vancouver locals as well as for a range of other audiences. Listeners suggested genres beyond musical composition for it, likening it to a soundscape story, oral documentary, and radio shows, though, as McCartney notes, the piece does not exactly conform to any of these formats (pp. 224–226). Westerkamp's reference to her own production process drew a laugh from every audience for whom McCartney played the piece, because it self-reflexively undermines its own rhetoric of objectivity as a documentary or as a sound recording (p. 230).

Several listeners also noted what Westerkamp would have called its schizophonic dimension, and Stanyek and Piekut would call its rhizophonic dimension: "'The view is beautiful,' No it's not. […] I don't know this space" (p. 232, see also pp. 234–236). For these auditors, spatial listening is ambivalent. When writing about the separation of sounds and sources, Westerkamp uses Schafer's schizophonia concept to argue that it is a problem when there is a difference between what is heard and seen in an environment because it has separated the listener and the environment (1988, pp. 26–34). Conversely, Stanyek and Piekut use the term *rhizophonia* to claim that *all* sounds are separated from their sources by definition (2010, p. 19). I am much more sympathetic to Stanyek and Piekut's interpretation—it's not a problem that Westerkamp is there and McCartney's auditor is not. But my goal is not to point out a contradiction between Westerkamp's writing and audio work and its reception in order to critique it. Rather, I highlight the contradiction in order to show that Westerkamp's audio work can be heard through entirely different hermeneutics from her writings; to different ends; and that such hermeneutics are equally valid.

Several of McCartney's listeners also noted the pedagogical dimensions of *Kits Beach Soundwalk*. This is no accident. Westerkamp's audio work is an extension of a well-defined pedagogical project in her writings and public lectures, from her classic 1974 soundwalking essay on down to more recent work (e.g., Westerkamp, 2007, 2010). And in turn that work extends and refines ideas in R. Murray Schafer's writings in *Ear Cleaning* (1967), *The Soundscape* (1994), and elsewhere. But that is precisely the point: *Kits Beach Soundwalk* is designed to actuate the mode of listening it wishes to inspire, and it does so through the sonic equivalent of diagrams and illustrations, as one might find in a book or presentation. In the visual domain, this is customary and common; in the audio domain, it is still rare to find audio examples next to the language describing them, and even notable digital humanities platforms like Scalar and Manifold still have video capabilities that far outstrip their affordances for working with audio. This is why it is useful to look back at Westerkamp's work as a precursor to the multimodal scholarship so central to digital humanities today. In *Kits Beach* and in her radio work, Westerkamp provides a model for using sound didactically, for teaching with it through media, and for disseminating work outside the normal academic channels.

At the end of his introduction, Rutherford-Johnson notes that Westerkamp's work challenges what is normally thought of as Western Art Music, showing that the borders of the term have become "highly permeable and fuzzy" (2017, p. 21).[5] But its status as scholarship should be uncontroversial: it is an exceptionally effective teaching tool in the modern sound studies classroom. For the past few years, in my large-lecture undergraduate sound studies course, I asked students to listen to *Kits Beach* as their first introduction to audio recording as an aesthetic practice (at least for many of them). I then ask them to work with clips from the recording in Audacity, a free audio editor, after some basic instruction from me. This is one of the best days of the term, as students who had never thought they could or should work with sound aesthetically take great delight in editing, manipulating, and transforming Westerkamp's words, sounds, and messages (they have also occasionally recorded me giving the Audacity instructions and mixed those in for fun as well). This sets up a unit on the theory of sound recording, starting with Stanyek and Piekut's essay. Because students have heard Westerkamp and worked with sound themselves, the phenomena described by Stanyek and Piekut—which read as words alone might seem highly abstract for someone who has no experience with sound recording and editing—become more concrete.

It also makes a difference that my undergraduate classes include, year after year, majority female members. That it is Westerkamp giving the instructions and providing the raw material is not accidental. Andra McCartney recalls that "I first heard Westerkamp's music in 1989 [...] it had a galvanic effect on me: I wanted to compose electroacoustic music so much that I went out, rented equipment, and began within days of that first listening" (1998, p. 6). The gendered dimensions are important here. As recent scholarship has amply documented, women and gender nonconforming people are still often heavily discouraged from thinking of themselves as entitled to, or capable of, working with sound as a material, using recording technology (Born & Devine, 2015). This *practical* knowledge of recording—made possible through the group audition and discussion of Westerkamp's piece and group experimentation with audio software—provides a firm basis for much deeper cultural readings of recording as a process, such as the Stanyek and Piekut piece referenced earlier, and critical writings on the voice, mediality, and power that follow in the course outline (for the 2018 version of the course, see: http://sterneworks.org/COMS350-W18.pdf). My integration of the piece into my course is built entirely on this reading of it. One can certainly appreciate *Kits Beach Soundwalk* as art, and that appears to be the dominant mode of writing on it, but I hope I have demonstrated it is also possible to appreciate the recording as pedagogy, and thereby to appreciate it as a precursor to digital humanities practices that are more common today.

Read against Westerkamp's description of her own radio practice, it is no coincidence that the vast majority of sonic-pedagogical work—work that is designed for and within a sound studies framework broadly conceived—exists as podcasts. "Radio-like" (Freire, 2008) in their format and styling, free to move beyond the commercial imperatives and constraints of radio, and serialized through RSS, sound studies podcasts have, decades later, taken up the mantle of sonic education through sound. Most notable in this field is the impressive collection of recordings made in connection with *Sounding Out!: The Sound Studies Blog*, which document a host of sonic practices, ways of sounding and listening, and an impressive, international, and intersectional array of sonic experiences unconstrained by a common format (https://soundstudiesblog.com/podcast/). *Phantom Power: Sounds About Sound* explores sonic arts and humanities through long-form interviews and high-concept composition and programming (http://phantompod.org/about-us-2/). Sound scholars are still not at the point where we can insert sounds in our printed texts like we can images, but that time will come.

In the meantime, the most successful alternative to silent textuality in sound studies is *Sounding Out!: The Sound Studies Blog*, which could be read alongside DH projects that refuse high dollar aesthetics and expensive infrastructure of digital humanities projects, such as the Minimal Computing (see https://go-dh.github.io/mincomp/). They use ready, off-the-shelf blogging software, borrow audio and video hosting from Soundcloud, YouTube, and other commercial platforms, and provide access to the field for people who otherwise might not have it. "Our move to combine craft production with a group experiment in digital community building came from a desire to push the rhetorical boundaries of sound studies and the sensory nature of 'writing' itself" (Trammell, Stoever, & Silva, 2018, p. 91). This is not to say that their solution is perfect—dependence on corporate platforms comes with significant compromises. But the long run of the blog shows what is possible, and it has done more than any publication organ in the field to expand its reach and provide access for emerging scholars, who most need access to publicity about their work.

Meanwhile, in our classrooms, we have reached the point where audio playback is possible and even sometimes easy. Our students can both record and play back the sonic world outside of our classes. The World Soundscape Project's commitment to pedagogy existed in a very different moment, where it was simply a radical thing to make one's own media and to separate one's sensibilities from those of the mainstream media. Our challenges today are different, but not too different. The tools may now be available, but increasingly, the major obstacles to their use are ideological and inertial. Westerkamp's patient, pedagogical voice is a model and a resource for teaching with and through sound.

Riverrun: Performing New Sonic Temporalities

"Barry Truax!"

These were the first two words spoken to me by Gerhard Behles, one of the inventors of the software program Ableton Live, and president of

Ableton. It was 2012, and I was visiting Ableton's headquarters, in an effort to understand the software's particular relationship to sonic time. For those who don't know it, Live is the Power Point of electronic music software[6]—ubiquitous in electronic music settings—because it provides musicians with a way of using their laptops as electronic instruments in real time for performance. One of Live's core features is time-stretching, which reprocesses audio clips recorded at different tempos so that they can play together at the same tempo, or proportionately matched tempos, without changing the pitch or other signature dimensions of the sound like envelopes and transients. This feature can also be used and abused for creative purposes, especially in more extreme settings (where envelopes and transients take on new characteristics). Originally, Live's time-stretching feature was built around granular sampling, a process Barry Truax helped popularize by engineering a way to do it in real time in the late 1980s.

Behles learned about Truax's work on real-time granular synthesis and sampling while a student at the Institute of Sonology at The Hague (author interview, 2012).[7] But Behles' greeting to me provides a window into a reception history of Truax's ideas about sonic time, as embodied in his computational and compositional practice. Behles described Live's time-stretching techniques as "fucking with time," by which he meant that time-stretching technology in Live—distantly derived from Truax's work (among others)—allowed artists and musicians to explore novel relationships between sound and time. When played back on an acoustic phonograph or through a standard pulse code modulation (PCM)-based digital playback system, the pitch of a sound recording will vary with its rate: faster playback leads to higher pitch; slower playback leads to lower pitch (Pohlmann, 2011; see also Feaster, 2011, for a discussion of the aesthetic implications of this phenomenon). Granular processing is one of a group of analog and digital technologies that does away with this relationship: sounds can be sped up or slowed down with the pitch maintained, or pitches can be shifted with no change in playback speed.[8] In the 1970s at the University of California Santa-Barbara, Curtis Roads found a way to realize this process in the digital domain. Barry Truax's work in the 1980s allowed this process to be done in real time and, therefore, to be subject to real-time control by a musician or a performer (Roads, 1988). This is why it was notable for Behles.

Whereas most sound synthesis had been built on theories of sound that decompose it into infinite waves of particular frequencies, granular synthesis and sampling treat sound as composed of particles that have a definite frequency-duration relationship. Drawing on theories advanced by Dennis Gabor (1946, 1947), Iannis Xenakis (1992), and others, granular processes compose audio from these "acoustic quanta"—minimal divisible units of sound (Roads, 2004).[9] Today, most of the processes called granular synthesis are, in Truax's nomenclature, species of *sampling* or *resynthesis*, where they take recordings of sounds as their basis for granulation, rather than generating the material basis for the granulation themselves. Truax developed a real-time approach to granular synthesis before he developed real-time granular sampling (see Roads, 2004), which is why *Riverrun* is built off synthesized sounds while later works are built off recordings—first phonemes in *Wings of Nike* (1987) and then eventually full environmental sounds, as in *Pacific* (1990). By 2000, real-time granulation on standard, consumer-grade computers with consumer sound cards was commercially available in programs like Pure Data (1996), Cycling74's Max/MSP (1997), Native Instruments Generator (1998), and Audio Ease's Riverrun (2000), the last clearly named for Truax's piece.

Over five major movements, played over four channels (a stereo version is also available), *Riverrun* ebbs and flows, from small clicks and bleeps to huge onrushes of sound that fill up the audible spectrum (see Bouchard, 2009).[10] To add some texture to this inadequate verbal description, on October 25, 2018, I conducted a listening session at the Sound Studies Listening Group at Northwestern University (SSLG, pronounced "slug"; organized by Jacob Smith), where I played a stereo version of *Riverrun* for an attentive audience of sound scholars and less "expert" listeners. As is their practice, I offered little context beforehand, though afterward, I explained my interest in the work. I did this to get outside my own head as a listener, in part because my own listening to the work was so motivated and shaped by the process of its production and Truax's (and critics') discourses around it, which might well fall into a discourse of authorial intent. Several striking things emerged from the conversation.[11] First, for a group used to thinking about sound but *not* necessarily familiar with granular processing as a sound production process, the people in the group immediately turned to a version of causal listening. The goal here was not to identify causes, as Michel Chion (1994) would have it, but to develop a vocabulary for describing what they hear through comparison—what it sounds *like*. Here people hear vastly different things: machinic and grating versus oceanic and flowing. This eventually coalesced into hearing the piece as working through the relationship between "the natural" and "the digital," but here the piece was generally heard in a romantic relationship to nature. This is in part based on the heard relationship between the sound and the name of the piece, taken from the first word in James Joyce's *Finnegans Wake*, which is itself a meditation on sleep and waking time, and therefore on consciousness. Prompted by a group of music theorists in the room, especially Mark Butler, we moved from these more direct connections to a consideration of form. There is very little silence in the piece—there are no conventional rests. This constant fullness over almost 20 minutes prompts a consideration of the spatiality of sound—as if one were "inside" it. Another set of contradictory hearings focused on embodiment versus disembodiment: some heard themselves being pulled out of their bodies; others heard a queering of sound, and an ambiguous, nonbinary gender performance, especially as sounds and movements blended into one another. Still others pointed to the different scales at which the piece operated, from the individual grains up to the entire structure. One listener called it "sneaky…it is teaching you to listen for resemblances."

Apart from that last comment, *Riverrun* is not didactic in the same way that *Kits Beach Soundwalk* is. It may be heard as trying to teach its auditors to listen differently, but it is doing so in the manner of many twentieth-century avant-garde musical works. There is no voice giving instructions, aurally pointing to a phenomenon and saying "Now listen to *this*." Yet, if we zoom out, *Riverrun* can be read as a meditation on the plasticity of sonic time and space because of the method of its composition and because of its particular sonic structure. Here, I want to consider the piece and the process behind it, granular synthesis, as an argument about sound. Truax himself made this argument in several forms. In "Capturing Musical Knowledge in Software Systems" (1991), he argues that musical software is a representation of ideas about sound. Traditional software for composers had been built around a literacy-based model of composition, as a replacement for the pen-paper-piano-score-reading-musicians assemblage that had been hallmarks of avant-garde composition for centuries. Crucially, Truax notes that technical limitations are not the key issue: "The strongest limitation is probably not physical or contextual constraints, but rather the individual's attitude toward compositional process" (p. 220). To this end, he viewed software instrumentally, using a stable but probably already-obsolete DMX-1000 system, as something that should be developed in order to produce "a critical mass of open-ended control possibilities for the user," today an axiomatic dimension of music software design, but in 1991 a somewhat radical proposition, especially in comparison to approaches to software at large computer music centers such as CCRMA at Stanford and IRCAM in Paris (see Born, 1995; Nelson, 2015).

Truax's system for *Riverrun* thus had a knowledge component, a particle theory of sound; a processing component, granular synthesis; and a praxeological component that enabled flexible, real-time control for the musician. Theoretically, granular synthesis challenged prevailing conceptions of pitch, time, and scale. "The basis of granular synthesis in the seemingly trivial grain has had a powerful effect on my way of thinking about sound. It clearly juxtaposes the micro- and macro- levels, as the richness of the latter lies in stark contrast to the insignificance of the former" (1990, p. 123). Sound synthesis had traditionally happened on a macro scale, based on sounds that corresponded to basic mathematical functions—most notably the sine wave—as the fundamental building blocks of sound. This was important for subtractive, additive, and frequency-modulation-based approaches to synthesis (see Nelson, 2015; Rodgers, 2011a, 2011b). Each of these approaches is in turn built around Josef Fourier's theory of heat as it was applied to sound in the nineteenth century (Hui, 2013; Kittler, 2017; Krämer, 2006; Pohlmann, 2011). A core characteristic of this math was that the waveform was theoretically infinite: it was defined purely in terms of its frequency, or how many times it cycles per second. But as Denis Gabor pointed out, sounds are not infinite: they have a definite duration (1947, p. 591, see also Gabor, 1946). Moreover, their duration affects the possibility for perceiving frequency both because of the physics involved and because of the limitations of human hearing (1947, p. 593).

Therefore, in contradistinction to the theory of sound as a continuous wave, which still pervades humanistic discussions of the materiality of sound, Gabor posited the existence of an acoustic quantum, the smallest unit of discernable sound (p. 592). It is a measurement of two quantities in relation to one another: frequency and duration. The result, as Curtis Roads has shown (2004, pp. 1–42), is that basic musical concepts like pitch and rhythm are challenged as absolutes: they only exist at one or another scale. Based on his reading of Gabor's critique of Fourier, Truax considered granular synthesis as one of a set of techniques that inaugurated "The End of the Fourier Era" (1992, p. 29). Among the ideas he hoped to push beyond are acoustic models that assumed perceptual absolutes (like the separation of pitch and rhythm), stimulus-response-based psychoacoustic models based on independent variables, models of composition drawn from the technics of literacy concepts like harmony and counterpoint, and notions of abstract, context-free art. Crucial here is that Truax's engagement with technologies, first at the Institute of Sonology in The Hague, and later in his own work on the DMX-1000, led him to these positions. They required engagement with the challenges of signal processing and the models of sound and music upon which it operated.

Other scholars have already well documented Truax's DMX-1000-based system for *Riverrun* (see Clarke, Dufeu, & Manning, 2014), but a few salient features are worth noting here. Although at the processing level, Truax talks about fairly traditional concepts for sound synthesis like attack and decay (envelope), central frequency and frequency range, and delay time (1988, p. 17), his system allows for their real-time production on (at that time) unprecedentedly small timescales: 10–20 milliseconds. Working with these tiny events required a different approach to controlling them, since thousands of grains might occur in a single second. Using a command-line interface, Truax developed a set of procedures: presets, which specify states of the various parameters of sound for each grain or groups of grains; ramps, which are "patterns of change in the parameters at a specific rate" that can be initiated at specific times by the user; and tendency masks, which combine presets and ramps but appear as graphical control shapes to the user "and hence suggest a different compositional approach." The key to using these techniques, writes Truax, is to abandon the hope for a deterministic result (the way an orchestra composer might predict the outcome of a score) and instead engage with the program's sound-generating capability *processurally* (pp. 19–20). Today, such an approach to signal processing software would be unremarkable—it is a standard feature of everything, from digital audio workstations, to software synthesizers, to apps, to audio hardware. But it was not the dominant paradigm of working with digital sound for composers in the 1980s, and Truax's mode of thought showed how approaching sound from a granular perspective implied a different kind of relationship to creativity, control, and sonority.

In combining theory, processing and an interface, Truax's approach to sound was not abnormal for computer music at the time (as a brief read of computer music publications would show). But it is worth meditating upon his specific approach because it proposed a radical alternative theory of sound and mediation—sound as a particle, rather than a wave—and because it did so through the process of building and using a technology and a set of techniques. Anne Balsamo's (2011) approach to "designing culture" argues that all technological artifacts are both technical and cultural and that technical experimentation thus also works as a form of cultural experimentation. Her concept of "hermeneutic reverse-engineering" is crucial here, because it combines the interpretative work usually associated with humanists ("hermeneutics") and the more mechanical study undertaken in engineering contexts to argue for a hybrid process (pp. 11, 14–16). Similarly, Matt Ratto uses the slightly awkward term "critical making" to signal a desire to "combine two modes of engagement with the world that are often held separate—critical thinking, typically understood as conceptually and linguistically based, and physical 'making,' goal-based material work" (2011, p. 253). As Ratto explains, the difference between his approach and the more traditional design- and engineering-based approaches to fabrication lies in that "Our main focus is on the act of shared construction itself as an activity and a site for enhancing and extending conceptual understandings of critical sociotechnical issues" (p. 254).

Obviously, Truax understood his project as primarily compositional, rather than humanistic or social scientific. Indeed, when I visited him at SFU in 2013, Truax told me he doesn't normally consider his work on acoustic ecology and his compositional and computational work to be connected. Yet they can clearly be read as intertwined, from the environmental themes of works like *Riverrun* and *Pacific* to the broader questions about the inner workings of sound raised by the technique of granular sampling and synthesis on which it is based, as well as his critique of composers' unwillingness to clearly and even didactically deal with real-world issues: "Unfortunately, people in computer music do not seem to have the same kind of broader concerns for social issues or the media, or if they do they don't see these are related to their professional work. For the most part composers seem wedded to abstract music [...]. Their work doesn't influence the environment and they don't let the environment influence their music" (Iwatake & Truax, 1994, p. 20; see also Truax, 1992, pp. 38–40). Truax's work with *Riverrun* led to the shaping of techniques and practices that would be central to his approach to sound recordings drawn from the real world. If one can hear the ebb and flow of a river through granulated sine wave and FM particles, one is prepared to hear the sonic worlds inside a single phoneme, or in recordings drawn from around Vancouver. Through its temporal expansions and superpositions, *Riverrun*—and the processing, theory, and parameters behind it—announces the pluritemporality of sound and then becomes a field of play for sound and time. Truax was not the first person to notice this possibility, not by decades. But he provided a vital and lasting framework to engage with sound and time as separable phenomena in real time, and a tool for others to use in their studies. That is what a good media theory does.

Conclusion: The WSP as an Archive of Sonic Practices

Decades before the term *digital humanities* was coined in 2001 (Kirschenbaum, 2010), and while English and other humanities departments were having debates about incorporating computers into their work, the World Soundscape Project was experimenting with combining humanistic modes of knowledge with media production and with alternative forms of knowledge dissemination. In this way, they were far ahead of their time and now provide an important sonic precursor to today's digital humanities. In mainstream academic publication and pedagogy, until recently presenting recorded sound with written scholarship was both difficult to achieve and rare. Books accompanied by compact discs were available from the 1990s, but the CDs were often lost, or, conversely, never played. They required too much coordination on the part of the user. Outside music schools, and sometimes in them, many classrooms had limited facilities for audio playback, an issue that surprisingly plagues us even today, as even sound-based panels and conferences sometimes have to work extra hard to have the most basic audio playback technology despite the ready availability of display technologies for slideware. There were novelties, like Marshall McLuhan's *Medium is the Massage* record, but these were exception that proved the rule. McLuhan's own involvement in the record was quite limited: he recorded his voice parts before creative director Jerome Agel and producer John Simon assembled the rest of the sound work.

Platforms that allow easy commingling of text and sound could and should have a profound effect on how we deal with sound in the humanities, now that it is more available to us in textual form. It is therefore useful to go back and think through how arguments have been rendered sonically in the past. Through my discussions of just two signature works from people involved with it, I have shown how the World Soundscape Project's body of sonic practice provides a rich repository of attempts to think sonically through sound and use audio as a tool for teaching theories of sound, culture, and media.

Long before the digital humanities came into focus as a field of practices, we can find people and practices who used analog media to accomplish some of the same things. One possible reading of Westerkamp's work would be rendered in terms of sound studies pride, to claim that "we were doing it first." But such a claim would have to leave aside traditions of filmmaking in visual anthropology, experiments with audio and video at the Media Education Foundation, and several other precursors. Instead, the takeaway should be that multimodal scholarship, teaching, and knowledge production are not so much about the technical tools the scholars use (despite the "multimodality" in the name) but rather the epistemologies and techniques of scholarly inquiry and argument that scholars can mobilize. The theoretical work of media production is most effective when, in beholding it, we focus more on the ideas and less on the tools or the methods that get us to those ideas, unless the theory is about the tools, as it was in *Kits Beach Soundwalk.*

Freeing ourselves to use new approaches and tools is a great and laudable goal, but we must also remember that new techniques of production always require new hermeneutics and appropriate modes of circulation and dissemination. This is common in writing but still rare in how humanists engage with multimodal work. Years of learning to read turn into comprehension of difficult texts; years of difficult reading and research lead us to advance new and challenging ideas. The same challenge now faces those scholars who are working in new modalities. If one of the goals of a fully realized sonic humanities is a fully realized sonic mode of scholarship, it will require that we sound in the idioms that our audiences are ready to hear or that we get our audiences ready to hear in the idioms in which we sound. Hildegard Westerkamp, Barry Truax, and even R. Murray Schafer understood this already in the 1970s, which is why they wrote, sounded, and taught in combination with one another. Their experiments are very much of their time and place, but they remain instructive for us today.

Notes

1. Thanks to the editors for the opportunity and their patience; to Carrie Rentschler, Andy Stuhl, and Rachel Bergmann for readings of a draft; to Mara Mills for many hours of rich conversation about sound, media, and time; to Tara McPherson for an opportunity to really think through sound and the digital humanities; to Neil Verma and Jake Smith and the SSLG at Northwestern University for a place to be heard; to Andra McCartney for

some helpful exchanges about the WSP; and to Barry Truax and Hildegard Westerkamp for talking with me and for their work.

2. At the risk of ruining the joke by explaining it for nonnative speakers, or for readers who stumble across this work in the future: "O.G." is a slang term from hip hop, short for "original gangsta," which refers to practice that is authentic, "old school," or the basis of something that others developed upon.
3. As of this writing, the piece is currently available to hear at https://www.youtube.com/watch?v=hg96nU6ltLk. It is also available on Westerkamp's *Transformations* CD.
4. This is not strictly correct. Andra McCartney (1999, p. 217) discusses score for the spoken part of the piece. NB: the digital copy of McCartney's dissertation lacks on-page pagination. Thus, I have cited the pages as displayed in a PDF reader so they are easier to find, even though they clearly do not conform to the original pagination for the paper document as suggested by the Table of Contents. I am using this copy: http://hildegardwesterkamp.ca/resources/PDFs/writings-pdf/Andradiss.pdf
5. From a definitional standpoint, Rutherford-Johnson is certainly correct, though from the standpoint of practice, the institutions of Western Art Music are still relatively *im*permeable for a wide swath of people and practices and remain overwhelmingly male (Born & Devine 2015). One hopes a new generation of composers, with ears for soundscapes, sound art, pop, funk, metal, and electronic music, might change this.
6. I am certain the company would hate this decidedly uncool comparison, but it illustrates Live's ubiquity in electronic music and sound art settings. In my experience, Live is a deeper and more engaging program in terms of its creative potential. That said, its "clip view" is sort of like an audio slide deck, where sounds can be played in sequence, in sync with one another, or—and this is where it far surpasses slideware—out of sequence.
7. I discuss the line from Truax's granular synthesis and sampling to modern software applications like Live more fully in a book I am currently co-writing with Mara Mills, entitled *Tuning Time: Histories of Sound and Speed*.
8. The earliest version of these technologies was developed in the 1930s and 1940s, using magnetic tape and optical sound-on-film technologies, so it is not an inherently digital process.
9. Wavelet synthesis works on an analogous principle (see, e.g., Kronland-Martinet, Morlet, & Grossman, 1987).
10. As of this writing, it is currently available online at https://www.youtube.com/watch?v=u81IGEFt7dM
11. This is my reconstruction of it from notes taken at the time, then turned into an account a few hours after the fact and edited here.

References

Akiyama, M. (2010). Transparent Listening: Soundscape Composition's Objects of Study. *RACAR, 35*(1), 54–62.

Balsamo, A. (2011). *Designing Culture: The Technological Imagination at Work*. Durham, NC: Duke University Press.

Blake, A. (2010). An Audible Sense of Order: Race, Fear and CB Radio on Los Angeles Freeways in the 1970s. In D. Suisman & S. Strasser (Eds.), *Sound in the Age of Mechanical Reproduction* (pp. 159–180). Philadelphia: University of Pennsylvania Press.

Born, G. (1995). *Rationalizing Culture: IRCAM, Boulez and the Institutionalization of the Musical Avant-Garde*. Berkeley, CA: University of California Press.

Born, G., & Devine, K. (2015). Music Technology, Gender and Class: Digitization, Educational and Social Change in Britain. *Twentieth-Century Music, 12*(2), 135–172.

Bouchard, A. (2009). *CEC — eContact! 11.2 — Granular Synthesis and Barry Truax's Riverrun*. Retrieved December 14, 2018, from https://econtact.ca/11_2/bouchard_truax.html

Brackett, D. (2016). *Categorizing Sound: Genre and Twentieth-Century Popular Music*. Berkeley, CA: University of California Press.

Chion, M. (1994). Audio-Vision: Sound on Screen. New York, NY: Columbia University Press.

Chun, W. H. K. (2011). *Programmed Visions: Software and Memory*. Cambridge, MA: MIT Press.

Clarke, M., Dufeu, F., & Manning, P. (2014, June). Barry Truax Riverrun (1986/2004), a Case Study for the Tacem Project, Exploring New Approaches to Techniques of Analysis and Re-Synthesis in the Study of Concert Electronic Works. *Proceedings Presented at the Electroacoustic Music Studies Network Conference*, Berlin. Retrieved from https://pdfs.semanticscholar.org/a33e/5a5686576788304aeeabb7f2c5565629a8ff.pdf

Curran, M. (2010–2018). *The Market*. https://themarket.blog

Deleuze, G. (2001). *Cinema 1: The Movement Image* (H. Tomlinson & B. Habberjam, Trans.). New York: Continuum.

Dyson, F. (2009). *Sounding New Media: Immersion and Embodiment in the Arts and Culture*. Berkeley, CA: University of California Press.

Eidsheim, N. S. (2015). *Sensing Sound: Singing and Listening as Vibrational Practice*. Durham, NC: Duke University Press.

Ernst, W. (2016). *Sonic Time Machines*. Amsterdam: University of Amerstdam Press.

Feaster, P. (2011). "A Compass of Extraordinary Range": The Forgotten Origins of Phonomanipulation. *ARSC Journal, 42*(2), 163–203.

Freire, A. M. (2008). Remediating Radio: Audio Streaming, Music Recommendation and the Discourse of Radioness. *Radio Journal, 5*(2–3), 97–112.

Gabor, D. (1946). Theory of Communication. *Journal of the Institute of Electrical Engineers, 93*(3), 429–441.

Gabor, D. (1947). Acoustical Quanta and the Theory of Hearing. *Nature, 159*, 591–594.

Galloway, A. (2012). *The Interface Effect*. Cambridge, UK: Polity.

Gitelman, L. (2006). *Always Already New: Media, History and the Data of Culture*. Cambridge, MA: MIT Press.

Gopinath, S. (2013). *The Ringtone Dialectic: Economy and Cultural Form*. Cambridge, MA: MIT Press.

Hayles, N. K. (2012). *How We Think: Digital Media and Contemporary Technogenesis*. Chicago: University of Chicago Press.

Hui, A. (2013). *The Psychophysical Ear: Musical Experiments, Experimental Sounds, 1840–1910*. Cambridge, MA: MIT Press.

Iwatake, T., & Truax, B. (1994). An Interview with Barry Truax. *Computer Music Journal, 18*(3), 17–24.

Kahn, D. (2013). *Earth Sound Earth Signal: Energies and Earth Magnitude in the Arts*. Berkeley, CA: University of California Press.

Kelly, C. (2009). *Cracked Media: The Sound of Malfunction*. Cambridge, MA: MIT Press.

Kim-Cohen, S. (2018, June). *Dark Optimism/Bright Pessimism: Listening Through Neoliberalism*. Presented at "Spectres de l'audible: Sound studies, cultures de l'écoute et arts sonores", Paris.

Kirschenbaum, M. (2010). What Is Digital Humanities and What's It Doing in English Departments. *ADE Bulletin, 150*, 55–61.

Kittler, F. (1999). *Gramophone-Film-Typewriter* (G. Winthrop-Young, Trans.). Stanford, CA: Stanford University Press.

Kittler, F. (2017). Real Time Analysis, Time Axis Manipulation. *Cultural Politics, 13*(1), 1–18.

Knouf, N. A. (2016). *How Noise Matters to Finance*. Minneapolis, MN: University of Minnesota Press.

Krämer, S. (2006). The Cultural Techniques of Time Axis Manipulation. *Theory, Culture and Society, 23*(7–8), 93–109.

Kronland-Martinet, R., Morlet, J., & Grossmann, A. (1987). Analysis of sound patterns through wavelet transforms. *International Journal of Pattern Recognition and Artificial Intelligence, 1*(2), 273–302. doi:10.1142/s0218001487000205.

Labelle, B. (2015). *Background Noise: Perspectives on Sound Art* (2nd ed.). New York: Continuum.

Leslie, M. (1970). In Defence of Anachronism. *Political Studies, 18*(4), 433–447.

Lingold, M. C., Mueller, D., & Trettien, W. (Eds.). (2018). *Digital Sound Studies.* Durham, NC: Duke University Press.

McCartney, A. (1998). Soundwalk in the Park with Hildegard Westerkamp. *Musicworks, 72,* 6–15.

McCartney, A. (1999). *Sounding Places: Situated Conversations Through the Soundscape Compositions of Hildegard Westerkamp.* PhD Dissertation, York University, Toronto.

McPherson, T. (2009). Introduction: Media Studies and the Digital Humanities. *Cinema Journal, 48*(2), 119–123.

McPherson, T. (2018). *Feminist in a Software Lab: Difference + Design.* Cambridge, MA: Harvard University Press.

Morris, J. W. (2015). *Selling Digital Music, Formatting Culture.* Berkeley, CA: University of California Press.

Nelson, A. J. (2015). *The Sound of Innovation: Stanford and the Computer Music Revolution.* Cambridge, MA: MIT Press.

Peirce, C. S. (1955). *Philosophical Writings of Peirce.* New York: Dover Publications.

Pohlmann, K. C. (2011). *Principles of digital audio.* New York: McGraw-Hill.

Radovac, L. (2015). Muting Dissent: New York City's Sound Device Ordinance and the Liberalization of the Public Sphere. *Radical History Review, 2015*(121), 32–50.

Ratto, M. (2011). Critical Making: Conceptual and Material Studies in Technology and Social Life. *The Information Society, 27*(4), 252–260.

Rée, J. (1991). The Vanity of Historicism. *New Literary History, 22*(4), 961–983.

Roads, C. (1988). Introduction to Granular Synthesis. *Computer Music Journal, 12*(2), 11–13.

Roads, C. (2004). *Microsound.* Cambridge, MA: MIT Press.

Rodgers, T. (2011a). "What, For Me, Constitutes Life in a Sound?": Electronic Sounds as Lively and Differentiated Individuals. *American Quarterly, 63*(3), 509–530.

Rodgers, T. (2011b). *Synthesizing Sound: Metaphor in Audio-Technical Discourse and Synthesis History.* PhD Dissertation, McGill University, Montreal.

Rutherford-Johnson, T. (2017). *Music After the Fall: Modern Composition and Culture Since 1989.* Berkeley, CA: University of California Press.

Schafer, R. M. (1967). *Ear Cleaning: Notes for an Experimental Music Course.* Toronto, Canada: Clark & Cruickshank.

Schafer, R. M. (1994). *The Soundscape: Our Sonic Environment and the Tuning of the World.* Rochester, VT: Destiny Books. Originally published as *The Tuning of the World.* New York, NY: Knopf, 1977.

Stanyek, J., & Piekut, B. (2010). Deadness: Technologies of the Intermundane. *The Drama Review, 54*(1), 14–38.

Sterne, J. (2003). *The Audible Past: Cultural Origins of Sound Reproduction.* Durham, NC: Duke University Press.

Sterne, J. (2012). *MP3: The Meaning of a Format*. Durham, NC: Duke University Press.

Sterne, J. (2015). The Example: Some Historical Considerations. In D. T. Goldberg & P. Svensson (Eds.), *Between the Humanities and the Digital* (pp. 17–33). Cambridge, MA: MIT Press.

Thompson, E. (2002). *The Soundscape of Modernity: Architectural Acoustics and the Culture of Listening in America 1900–1930*. Cambridge, MA: MIT Press.

Trammell, A., Stoever, J. L., & Silva, L. (2018). The Pleasure (Is) Principle: Sounding Out! and the Digitizing of Community. In M. C. Lingold, D. Mueller, & W. Trettien (Eds.), *Digital Sound Studies* (pp. 83–119). Durham, NC: Duke University Press.

Truax, B. (1988). Real-Time Granular Synthesis with a Digital Signal Processor. *Computer Music Journal, 12*(2), 14–26.

Truax, B. (1990). Composing with Real-Time Granular Sound. *Perspectives of New Music, 28*(2), 120–134.

Truax, B. (1991). Capturing Musical Knowledge in Software Systems. *Interface, 20*(3–4), 217–233.

Truax, B. (1992). Musical Creativity and Complexity at the Threshold of the 21st-Century. *Interface-Journal of New Music Research, 21*(1), 29–42.

Westerkamp, H. (1988). *Listening and Soundmaking: A Case Study of Music-As-Environment*. MA Thesis, Simon Fraser University, Vancouver.

Westerkamp, H. (1989). Kits Beach Soundwalk (audio recording). *Transformations*. Montreal: Empreintes DIGITALes, IMED 9631/1996.

Westerkamp, H. (1994). The Soundscape on Radio. In D. Augaitis & D. Lander (Eds.), *Radio Rethink: Art, Sound and Transmission* (pp. 87–94). Banff, AB: Walter Phillips Gallery.

Westerkamp, H. (2007). Soundwalking. In A. Carlyle (Ed.), *Autumn Leaves: Sound and the Environment in Artistic Practice*. Paris: Double Entendre. Retrieved from https://www.sfu.ca/~westerka/writings%20page/articles%20pages/soundwalking.html

Westerkamp, H. (2010). *What's in a Soundwalk?* Public Lecture Presented at the Sonic Acts XIII, Amsterdam.

White, H. (1966). The Burden of History. *History and Theory, 5*(2), 111–134.

Xenakis, I. (1992). *Formalized Music: Thought and Mathematics in Music* (Rev. ed.). Hillsdale, NY: Pendragon Press.

SECTION II

Environment and Community

CHAPTER 6

Nothing Connects Us but Imagined Sound

Mitchell Akiyama

Step through your door and close your eyes. Listen. How do you know you are home? Are there any unmistakable sonic events or textures that ground you in this place? The electronic chime of a school bell at recess? The seasonal swish of a stand of poplar trees? The low hum of an electrical transformer? How singular is your sonic environment? Now step inside. What sounds fill your home? A radio? A podcast beamed via Bluetooth to a nearby speaker? The murmur of the neighbor's television? What worlds do these mediated, electrical sounds connect you to? Do these transmissions link you with local spaces or concerns? Or, is their reach national? Global? How does sound—acoustic or electronic—contribute to your sense of place, your sense of belonging?

In his 1977 book, *The Tuning of the World*, R. Murray Schafer made the emphatic case that sound fundamentally shapes community. Prior to the advent of electrical technologies capable of amplifying, storing, and transmitting sound, communities were defined by the unique sounds and acoustics of their immediate environments. Schafer invoked, for example, Plato's ideal community, which (as the apocryphal claim goes) should be limited to 5040 individuals—this apparently being the precise number of

M. Akiyama (✉)
University of Toronto, Toronto, ON, Canada
e-mail: mitchell.akiyama@daniels.utoronto.ca

M. Droumeva, R. Jordan (eds.), *Sound, Media, Ecology*,
Palgrave Studies in Audio-Visual Culture,
https://doi.org/10.1007/978-3-030-16569-7_6

souls an unamplified orator might comfortably address.[1] The sound of a church bell or the song of the local muezzin intoning the call to prayer; the geography of community was laid out along sonic lines. But the rise of industrial noise, Schafer argued, had come to drown out these sounds and sonic reproduction technologies had confused matters further by separating "original" sounds from their sources. Schafer blamed the industrial revolution for degrading the soundscape into a condition of low fidelity. In his terms, a "lo-fi" soundscape was characterized by consistent, loud noises emanating from industrial machinery, combustion engines, and myriad other mechanical drones. He contrasted this with the "hi-fi" soundscape—the quiet rural backdrop in which even relatively quiet sounds had space to assert themselves (1977b, p. 43).

In the late 1960s, at Simon Fraser University in Vancouver, Schafer founded the World Soundscape Project (WSP) to address these issues. The WSP emerged as a collective of young composers whose goal was the elimination of noise pollution and the engineering of a healthy, balanced soundscape. The WSP developed an array of pedagogical methods for teaching people to open their ears to a world of sound, a world to which they had become troublingly inured. These techniques included exercises such as guided soundwalks and writing detailed description of sonic events as they unfolded.[2] A key part of the repertoire they called "ear cleaning" was an engagement with sound recording. Working with recording was essential in that, according to Schafer, the development and misuse of recording technologies was a problematic contributor to the increasing noisiness of soundscape. Recording engendered a condition of psychic distress that Schafer dubbed "schizophonia" because it separated sounds from their sources (1977b, p. 90). But, paradoxically, Schafer also had faith that schizophonic media could help repair the social damage they had caused by reconnecting people to significant sounds that they stood to lose.[3]

This possibility, that a schizophonic medium might reimmerse people in a shared aural heritage, underpinned the WSP's ambitious, sprawling radio program, *Soundscapes of Canada*, a ten-part series that aired in 1974 on the Canadian Broadcasting Corporation's (CBC) venerable program, *Ideas.* Consisting of material gathered on a cross-country trip that spanned the nation's distant coasts, *Soundscapes of Canada* was an eclectic work that ranged from instructional lectures by Schafer on how to become a more sensitive, critical listener to extended and unedited environmental recordings to catalogs of significant Canadian sounds to experimental tape pieces. It was a revolutionary work of composition, but the social and political positions that the series promoted were, in contrast to the its aesthetics, tacitly conservative. The soundscapes they included in the broadcast overwhelmingly represented the nation's settler colonial past—simpler, quieter times they seemingly wished to restore. However, created at a time when Canada's immigrant populations were exploding, when indigenous activists were making important strides toward state recognition, *Soundscapes of Canada* was perhaps most notable for whom and what it left out.

Twenty years after Schafer's first efforts to draw attention to the fraught relationship between sound and community, a collective of activist-artists, known as Ultra-red, mounted their own effort to turn sound recording toward political ends. Equally unflinching and polemical, Ultra-red's approach to sound and politics was radically different from that of the WSP. The collective was composed of community organizers and members of the Los Angeles chapter of ACT UP who, beginning in the mid-1990s, began to sonically document the fallout of the AIDS crisis and its intersection with the social pressures of discriminatory housing policies, migration, and addiction. While the WSP was committed to a politics of media predicated on preservation, Ultra-red treated sound recording as an epiphenomenon of class struggle. While the WSP sought to capture the world as it was in order to use its recordings to help compose and construct its ideal soundscape, Ultra-red proposed an inversion of this formula, arguing that sonic media could not be relied on to provide stable or unproblematic representations. Instead, they dealt with microphones and recorders as tools that might help produce social spaces radically different from the ones they found oppressive. For Ultra-red, the content of a recording was never nearly as important as the very effect that the act of extending a microphone might engender:

> The microphone does not have a perspective on the site of struggle. It does not stand apart from the struggle. Rather, it is a site for the production of the conditions of struggle. Inquiry is conditioned by the collective organizing of demands. In a militant sound investigation, we take time to organize the social field to be recorded. (2008c)

Organizing the field in advance of the recording fundamentally alters the status quo of causality; the object of the recording is no longer the world as it is but the world as the recordists require want it to be. By not only recognizing themselves as intervening in, but also constructing the terms of representation, the recording would lose its value as a window onto events-as-they-occurred. For Ultra-red, the schizophonic rupture between sound and source offered an *opening* for new social formations and political possibilities: "Only by artifice can we even conceptualize urban space as distinguishable from its ambience. Separating sound from context produces the most artificial results: a utopia so to speak" (2008b). Ultra-red saw the slipperiness of sonic mediation as an opportunity to amplify difference, to admit marginalized voices into the contested, fractured spaces of political action. The WSP's *Soundscapes of Canada*, on the other hand, evinced concern for a perceived erosion of national identity at a time of intense social upheaval and took on the task of reversing, or at least slowing, the disappearance of the Canadian, settler soundscape.

But here we should press pause to note that these projects, conceived and received twenty years apart, were the products of very different social and technological moments. Both groups used microphones and portable recording equipment to capture sound in situ, but they both had very different senses of what those recordings contained and to what ends they might be used. These different approaches to recording remind us that technology, mediation, and representation are always politically resonant; they are always profoundly entangled with the social forces and debates they capture and represent. The WSP's output of the 1970s hewed to a logic of analog storage and transmission, a tacit ontology that subtended the tape recordings and radio transmissions they believed held the power to repair the soundscape. "Analog," as we have come to understand the term, denotes the process of a continuous signal being inscribed to a medium that registers the infinite subtleties and modulations of things such as they are (Robinson, 2008). The WSP was concerned with the ways in which something truthful, meaningful, and specific abided in their tape recordings—even if tape made radical transformations possible, even if sounds were fundamentally corrupted by their removal from their origins. While two episodes of *Soundscapes of Canada* included compositions that featured radical transformations of the source material (episode IX: "A Radio Programme About Radio," for example, was a lysergic, almost Dada-esque meta-commentary on mediation), the majority of the broadcast eschewed any significant manipulation of the recordings made by members Peter Huse and Bruce Davis on their epic cross-country field trip. The WSP's choice to leave the majority of their recordings as they were suggests that they understood that there was something important about the connection between a sonic reproduction and its source—in spite of their anxieties about the medium's fundamentally schizophonic nature. It bears remembering that *Soundscapes of Canada* was created at a moment when the tide of postmodernism was barely lapping at the shores of culture, a moment in which modernist edifices were starting to quake, but had not yet collapsed. Grand narratives still held, and media storage still captured the stream of the real whose flow would eventually be fragmented by digital encoding.[4]

Ultra-red was no less tethered to the social and technological contingencies of the 1990s. In its fragmented, agonistic approach to politics and aesthetics, their work is emblematic of a digital postmodernity defined just as much by the arbitrariness of numerical encoding as it is by the slipperiness of language and narrative. Media, in their various regimes, fundamentally affect how we represent the world and marshal political agency within it. Yet also, at the same time, the culture and discourse of a moment will always inform technological possibility. Listening in on two distinct networks of technology and networks of discourse that open, close, and bleed between the 1970s and the 1990s, we will hear the echoes of a world roiling and shifting between modes of mediation and political desire.

The comedown following the national giddiness of Expo 67 in Montreal was harsh. Political assassinations and the threat of separation in Quebec, economic collapse, northern lakes left lifeless on account of acid rain. Canada's perennial grandfather-historian, Pierre Berton, would later longingly describe 1967 as "the last good year" (1997). The cultural climate of the early 1970s crackled with charge and change. But it wasn't only the political landscape that rippled with turbulence; the soundscape had also become a mess. The WSP, whether in response to the zeitgeist or not, took up the challenge of shoring up the country's troubled, shifting identity by recording its people and places, by playing the nation back to itself.[5] Their first foray into composing with recordings came in the 1973 work, *The Vancouver Soundscape*. But it was *Soundscapes of Canada*, which aired the following year, that articulated the group's ambitious goal of restoring the integrity of the nation's sonic environment. One episode in particular, "Soundmarks," stands out for its attempt to both define and foster sonic community. In the program's introduction, Schafer defined the soundmark as "a special feature of a community that helps to give it its unique character" (Schafer, 1974). Schafer (1974) introduced the "Soundmarks" episode as the result of "an extensive tour across the country, from Newfoundland to British Columbia." But "extensive," in this case, did not necessarily mean thorough or complete.

Created at a time when Canada was coming into its own as one of the most multicultural nations in the world, "Soundmarks"—and the entire series more broadly—was significantly lacking in representations of the country's exploding diversity. The sixth episode in the series offered an hour-long aural catalog of church bells, foghorns, train whistles—all symbols of the nation's settler colonialist heritage. For representations of the myriad non-white, non-settler soundmarks, Canada's Others would have to listen elsewhere. This is not to suggest that this occlusion was deliberate or the premeditated enactment of a specific political ideology. Four decades later, Bruce Davis (who, along with Peter Huse, made the recordings that were the basis of *Soundscapes of Canada*) admitted that their methodology for selecting subjects and locations was based largely on instinct.[6] It never occurred to them to use quotas to guide their fieldwork. Their motivations seemed to be based on good faith and a curiosity about the national soundscape, but we cannot lose track of the always-real discrepancy between the creators' objectives and an audience's reception of their work. Imagining back to 1974, giving these young composers some benefit of some doubt as to their good intentions, it is still important to ask: what did these broadcasts mean to a Japanese-Canadian woman whose family had been interned in the interior of British Columbia during WWII, whose possessions had been seized by the government only thirty years earlier? To a Trinidadian-Canadian, just arrived in Toronto, trying to settle herself in a settler's landscape, her skin and her accent both dead giveaways of otherness. To a Cree boy in Churchill, Manitoba, whose grandparents had spirited just enough of their language into his ear for their endangered tongue to endure. It is important to ask these questions, because, in 1974, Canada contained ever more multitudes, and it was at this very moment that it was becoming less and less tenable to ignore the nation's swelling, multicultural population.

While the WSP had not fully worked through the political implications of environmental recording in the 1970s, their output of this era palpated with the social concerns of the day. Although he certainly did not always speak for the entire group, R. Murray Schafer's writings about race and Canadian identity offer important insights into the ideological underpinnings of the WSP's work. Schafer's writings, particularly those that deal with Canada, invariably circle back to the role that nature plays in forming a national identity. For Schafer, Canada's climate and its landscape were its richest cultural resources, the sources of its people's strength and identity. In the program notes to his 1973 composition, *North/White*, he submitted that "the Canadian climate and geography…is our best unifier, transcending ethnic extraction or allegiance of any other kind. We are all Northerners, sharing a million acres of wildness in the imagination. That is our only uncounterfeit resource, and we should seek to draw more directly from it" (1984, p. x). But, ever since the country's founding, the national identity that Schafer believed superseded ethnicity has never been anything more than an idealized composite. He was certainly not alone in assuming that rugged northern climate would eventually chisel a hearty Nordic subject out of even the most unfit prospects; this was indeed the very crux of the one-hundred years of discourse that preceded him (1984, p. 64). "The idea of North," wrote Schafer (1984) in his notes to *North/White*, "is a Canadian Myth. Without a myth a nation dies" (p. 64). This last sentence would be recycled, nearly verbatim, a year later in *Soundscapes of Canada*. Schafer's idea of North would not figure prominently in the radio series, but it would tacitly form the basis of a racial politics underpinned by a deeply troubled and troubling colonial legacy.

Since the very emergence of the modern nation-state, media have profoundly influenced national conversations about identity and belonging. Benedict Anderson famously argued that the rise of mass-produced broadsheet newspapers, maps, and censuses prompted people separated by sometimes vast geographic distances to imagine themselves as participating in a common political project. Anderson proposed that community has also historically been imagined and ratified through sound. In his consideration of national anthems, Anderson (2006) coined the term "unisonance" to describe the power that sound can have to seemingly erode the boundaries between self and other: "How selfless this unisonance feels! If we are aware that others are singing these songs precisely when and as we are, we have no idea who they may be, or even where, out of earshot, they are singing. Nothing connects us all but imagined sound" (p. 145). While Anderson was specifically addressing the power of song to marshal nationalist affects, we can certainly extend unisonance to other forms of sonic identification. The soundmark is a particular and concrete instance of this phenomenon, one whose immediate ties to nationalism are not quite so transparent. Soundmarks are local by definition, but in broadcasting them to the nation over state radio, the WSP called on sound recording to serve as a receptacle for an imagined, shared, national identity.

This gesture could only be effective if the listener trusted that a recording is truthful and *contains* a real trace of a specific place captured at a specific moment in time. In a sense, this is the case for all recordings, but there is an aesthetic of authenticity and truth that is amplified when recordings are understood as having been captured in the field. There is no obvious or metaphysical difference between a field recording and its antipode, a recording made in a studio. The difference—which is fundamentally rhetorical—is in the way that space is framed by the recording. The studio recording is meant to sound as though it is emerging directly from your speakers. It is a sonic event seemingly divorced from the site of its creation. A field recording, in contrast, seems to fully capture and represent a space in its plenitude at a particular moment in time. Listening to a field recording feels like passing through a window into another time and place. But it is a condition of genre, not ontology, that inflects how we receive a recording. It is a matter that has important consequences for what we understand a recording to "contain" and, consequently, helps to determine what one might do with it. Field recording is effectively a genre of practice that, while always subject to arbitrary decision-making and technical limitation, has been accorded a documentary and preservationist function.

While the term "field recording" had been in use for decades by the 1970s, the phrase is notably absent, save for a few passing references, from the WSP's writings from that period.[7] Perhaps this was because the members of the WSP came primarily from musical backgrounds and not from the social sciences, from disciplines such as anthropology and ethnology, where the term circulated as a matter of convention. In a sense, the WSP's use of the word "environment" stood in for what practitioners in other areas called "the field." And while they gave very little explicit attention to the idea of location—an omission that is surprising considering their rigorous (albeit unorthodox) documentation practices—questions concerning the veracity and authenticity of certain situations crop up again and again. It seems that they almost took it for granted that their listeners would understand what these recordings were and how they were made—that they *were* field recordings, plain and simple. The site-specificity of their recordings was underwritten by the narrative conceit of the field trip, of journeying to other places in order to represent them as they actually are. Given that the WSP never really articulated an approach to field recording in any cogent way, we're left to cobble a philosophy together out of scraps, such as this telling dialogue between Bruce Davis and Peter Huse. In Davis's words, "Whenever you record a sound, you're ripping it out of its social, historical, and general acoustic context, so that the difference, for instance, between our recordings and a sound effects recording is that the sound effects recording is just the sound, and our recordings are not only the sound, but also the related background material to that sound" (Davis & Huse, 1974, p. 32). For Davis, the difference between a field recording and a sound effect was a width of context, a responsibility toward the totality of a sonic environment.

They apparently wanted to have things both ways, to capture the world both as it was *and* as they wanted it to be. They were explicit about this, even in spite of their concern for letting recorded sounds be themselves, making no apology for intervening in a recording or choosing an opportune or token moment. Describing the WSP's pedagogical approach to sound recording, Schafer (1977b) wrote, "We train students in soundscape recording by giving them specific sounds to record…It is not easy if the result is to be 'clean,' without distracting interferences" (p. 210). This essentially reads as a coded protocol for omission. To be fair, the goal of this assignment was to capture specific sounds in isolation, but the lingering and concomitant insinuation remains that some sounds are more valuable than others. Schafer gave an example of such interference, describing a scenario in which a young boy ruined a recording of a noon whistle by asking the crew if the sound he hears is the one they were after (1977b, p. 210). The takeaway?. "One of the recordist's biggest problems," Schafer grumbled, "is to devise ways of recording social settings without interrupting them" (p. 210). But what of the social dynamics established by the recording apparatus itself? Paradoxically, it seems that "reality" could only ever be approached through the *artifice* of framing, if not outright staging. If capturing the related background material was what differentiated a soundscape recording from a sound effect, then it bears asking just which sounds were extraneous? And, if context was so important, then what are we to make of their mandate to record singular, historically important sounds in isolation? Schafer hoped that a future cohort trained by the WSP would embark on a mission to record rapidly vanishing sound objects—kerosene lamps, school hand bells, water pumps, leather saddle bags (1977b, p. 209), for example—before their extinction. But would this not amount to a catalog of sound effects ripped from their context of use? Every act of recording is an act of framing; every act of framing is an act of omission. The question is, how are these conditions acknowledged by the recordist?

It's important to pause again to note that the WSP's approach to the act of capturing sound and working with recordings developed over the following decades to include different methods and concerns.[8] In his 1984 book, *Acoustic Communication*, Barry Truax attempted to formalize the practice that he now called "soundscape composition" that had emerged from the WSP's work of the early 1970s. In "soundscape composition...it is precisely the environmental context that is preserved, enhanced, and exploited by the composer" (Truax, 1984, p. 207). Soundscape composition, for Truax, was more than a formalist exercise. As with other ear cleaning techniques, it was meant to change the ways in which people related to the acoustic environment: "...the successful soundscape composition has the effect of changing the listener's awareness and attitudes toward the soundscape, and thereby changing the listener's relationship to it. The aim of the composition is therefore social and political, as well as artistic" (Truax, 1984, p. 207). In order for a change to take place in the listener, it was crucial that the listener identify with the sound source, recognizing and identifying its naturalistic and referential properties. Hildegard Westerkamp would propose a similar definition, suggesting that soundscape composition held the power to change listeners' attitudes toward the environment. Perhaps more than any other WSP member, Westerkamp was sensitive to the paradoxes of schizophonia. She proposed that, on the one hand, "In soundscape composition the artist seeks to discover the sonic/musical essence contained within the recordings and thus within the place and time where it was recorded" (2002, p. 52). But she was also skeptical that recording could ever have any claim to objectivity insofar as she was mindful that soundscape recordings needed to be shaped and transformed by the composer for them to realize their full, affective potential (Westerkamp, 2002).

Soundscapes of Canada was produced at a pivotal cultural moment, but it also arrived right as the inexorable wave of digital media was gathering force. As with just about all the WSP's positions on recording technologies, their views on both analog tape and digital recording were deeply ambivalent. In *The Tuning of the World*, Schafer flip-flopped on the value of analog media, noting that, on the one hand, records and tape recordings not only disrupted the natural order by separating sounds from sources, but they also sowed cultural confusion: "A record or tape collection may contain items from widely diverse cultures and historical periods in what would seem, to a person from any century but our own, a meaningless and surrealistic juxtaposition" (p. 90). However, the tape recorder could also serve as "a useful adjunct to the ear," a supplement to hearing capable of capturing sonic events that the human auditory apparatus was liable to miss (Schafer, 1977a, p. 208). There was a similar ambivalence regarding digital recording in Barry Truax's writings from the years immediately following the publication of *The Tuning of the World*. In his 1978 publication, *Handbook for Acoustic Ecology*, Truax extolled then nascent digital recording technology for its ability to produce copies without the introduction of noise. Twenty years on, in a 1996 article revisiting the work of the WSP at a moment in which digital recording and production had become commonplace, Truax lobbied to restore an ethic of indexicality or transparency to composition. Truax argued that composers had squandered the potential of digital technologies by using them to simply appropriate the richness of recorded sonic environments back into a formalist, musical language—a lost opportunity to harness the timbral complexities of environmental sound, along with its referential properties (Truax, 1996, p. 49). What was at stake in achieving such a balance was "the re-integration of the listener with the environment in a balanced ecological relationship" (p. 63). But just who was this listener, and into which environment she or he might be reintegrated? These are important questions to ask, especially given the tacit prioritization of settler colonial identities in *Soundscapes of Canada*. By the mid-1990s, Truax's thinking seemed to shift significantly, and the answer to this question was no longer clear. Truax (1996) imagined prying open the previously sovereign, asignifying space of the composition so that it might admit including details that would connect it to the social world in which it would be received, details such as the age, gender, race, or class of the performer or composer.

Around the same time that Truax was trying to reconcile the contradictions of digital media, Ultra-red set out, a Digital Audio Tape (DAT) recorder in hand, to record marginalized members of their community—intravenous drug users, immigrants, the queer cruisers of Los Angeles parks. They would then transform these field recordings through digital processes, sometimes altering them beyond recognition. Ultra-red's compositions of the mid-late 1990s oscillate dizzyingly between naturalistic documentation and digital fragmentation. They might establish a scene only to tear the whole thing to glitchy pieces. Take, for example, the track, "Public Address (C. B.)," from their 1999 album, *Second Nature*. A voice, ostensibly recorded in a studio, describes the persecution faced by gay men for engaging in public sex in Los Angeles's Griffith Park. A few seconds in, the recorded voice is shattered into shards of digital sound, only to be reconstituted in what seems to be an outdoor locale. But the recorded voice continues to change; it becomes apparent that it is now emerging from a speaker that is being moved around what we'd assume is Griffith

Park itself. Spaces nested within other spaces, each of which could be shredded beyond recognition at any moment. There is no trusting in the truth of a document that can so easily be degraded, whose material and numerical fragility are so easily exposed.[9]

However, even this focus on the (non)veracity of recording misses the point. Ultra-red has consistently maintained that their recordings are not meant to function as documents; they are only useful to the extent that they make things happen. In a 2004 interview, Dont Rhine explained, "Personally, I'm quite skeptical about the potential for someone's consciousness to be raised by simply listening to a song, reading a book, or watching a movie…Consciousness is radicalized in the direct participation in struggle" (Macdonald, 2004, p. 16). Put another way, a recording, like a concept (to transpose a phrase from Brian Massumi) is a brick. Build a wall or lob it through a window; its power lies in its use (Massumi, 1987, p. xii). Ultra-red has been unequivocal about taking matters into its own hands, treating sound recording as a tactical fulcrum for prying open spaces of engagement. "A political aesthetic of field recordings, as we've come to understand it," Rhine noted, "is not organized around the truth of the record…Instead, the field recording tests our memories against the truth of our desires…The political site takes shape the moment those actors involved in the event listen to the recording and reflect on it in relation to their memories and desires. The record is never the same as we remember it to be" (Macdonald, 2004, p. 16). Electronically mediated sound gains its power not through representation, but through the ways in which it defamiliarizes, constitutes, and creates public spaces that sound and feel different from those envisioned by the experts—the architects, the planners, and technocrats—who have the mandate to build. Concrete architecture and *musique concrète*—each consolidates, reduces a manifold to a singular, material encounter. The question, then, is how does one open both built space and the sound object to the world in all of its complexity? And how does one bend the world away from what it is toward what it might be?

There is a self-avowed, utopian striving implicit in this understanding of mediation. The recording does not capture and contain a given time and place; it is a non-space shot through with ontological uncertainty. Politics, like nature, abhors a vacuum and something new will always rush in to fill the void. Fredric Jameson describes this motive, oscillating force as "the dialectic of Identity and Difference…a politics [that] aims at imagining, and sometimes even at realizing, a system radically different from this one" (Jameson, 2005, p. xii). Radical difference, not teleological striving; utopianism does not necessarily have to have a blueprint for the future. Ultra-red's utopian strategy explicitly leverages this flicker between identity and difference, between realism and artifice, that recording cracks open: "Only by artifice can we even conceptualize urban space as distinguishable from its ambience. Separating sound from context produces the most artificial results: a utopia so to speak…the artifice we construct gives shape to our own position in public space" (Ultra-red, 2008b). The separation of sound from source does not degrade some presumably integral and pure original; rather, it creates new experiences and affects that can remake the field itself.

This concern with identity and difference is always roiling within in Ultra-red's work in terms of both politics and mediation. But even this tension implies or upholds an ontological distinction between recording technologies and politics. Is it sufficient to simply say that media are always charged with political energy while politics are always affected by media? Maybe it is more accurate to say that they in fact constitute each other, that the technological possibilities of a moment actually afford particular social manifestations, just as the emergent discourses of a moment subtend our conceptions as to the political potential of media. An injunction, then, is to think media and politics not as a dyad or a dialectic, but as an entanglement of co-constituting material, social, conceptual, and metaphorical elements. Let us consider our givens: a collective of activists/artists stepping out onto the shaky, postmodern ground of 1990s; finding their footing amidst the tumult and vibration of identity politics; steeped in Deleuze, Attali, and Mouffe; straddling a tectonic shift hastened by digital media. In the 1990s, identity has become increasingly understood as a set of intersectional variables, all of which are, to some extent, effects of history and performance.[10] A similar characteristic is, of course, accorded to the logic of digital media as well. Digital information no longer seems to have any important ontological ties to its material substrate; our encounter with the digital always involves a transcoding from bits to electricity to light or acoustic vibration (Manovich, 2001). Any of these manifestations could just as easily be expressed in another physical register. As such, information has come to be understood as subject to the violent glitches and ruptures that make media and identity irreconcilable and recalcitrant.

Another twenty years on. We are more digital than ever, digital to the point that we require physical or natural metaphors just to keep our heads in the Cloud. It seems fitting, maybe more than coincidence, that in 2016, "fake news" swayed an election, while at the same time MIT researchers were developing an algorithm capable of analyzing video and then producing audio realistic enough to fool humans into believing both sound and image were captured simultaneously (Conner-Simons, 2016). Moving into a future of convincing, digitally simulated media production, we will likely find it harder and harder to trust the veracity of recordings. The advent of digital storage and processing has made it possible to alter and produce audio in ways that were unimaginable in the analog era.

One proposal for a temporary and imperfect fix for this crisis of fragmentation has been the careful deployment of what Gayatri Spivak, in the 1980s, presciently named "strategic essentialism" (1996, p. 204). Spivak, ever-always distrustful of essences, submitted that it might be politically expedient for oppressed peoples to rally around a common cause, a mission anchored by a shared but provisional identity. This notion has been resurgent in recent years, touted by no less a critic of essences than Bruno Latour, who sees this as a problematic but necessary move for combatting the idea that facts, such as those relating to climate change, are open to interpretation.[11] What, then, would a strategically essentialist approach to our fractured political situation *sound* like? Politics and media are now far too complicated and mistrusted to imagine that a sound recording could ever unify a people under the sign of a shared history such as the one offered by the WSP in the 1970s. And yet, their work can still offer important and productive insights into the powerful affects that shared sonic experience can afford. We might organize around the "content" of a recording as long as we are attentive to whom or what might be missing and remember all the while that a recording can never be anything more than a placeholder for the concerns it draws together. Given the looming likelihood that real-seeming sound will soon stream into all corners of our mediascape, it is critical that we understand that digital audio can never be trusted to deliver a "true" representation of the world to us. This does not mean that what seems to be there, enunciating itself audibly—a voice, the wind whispering through a stand of oak, the din of a construction site—is necessarily nothing more than a digital fiction or fabrication, nor is it without value to attend to the content of the recording. However, the meaning of a recording can only ever be articulated by the texts and discourses through which it is interpreted. What seems to be there might never even have been. But, then, the future will certainly come to be; better to organize around what a recording proposes than what it purportedly contains.

Notes

1. I call this claim apocryphal given that Plato's *Laws*—the apparent source of the figure, 5040—makes no mention of orators or the limits of spoken address. For Plato, the number 5040 was mathematically sacred for its versatility as a divisor, and for its frequent appearance in real-world and mathematically ideal phenomena. It is possible that Schafer was pointing to an uncited reference made by Lewis Mumford in *The Story of Utopias*. See Plato, *Laws*, The Loeb Classical Library (Cambridge: Harvard University Press, 1926); Lewis Mumford, *The Story of Utopias* (New York: Boni and Liveright, 1922), 39; R. Murray Schafer, *The Tuning of the World* (New York: Knopf, 1977b), 215.
2. See *Ear Cleaning: Notes for an Experimental Music Course* (Toronto: Clark & Cruickshank, 1967).
3. Bernard Stiegler describes this tension in media between danger and redemption by invoking the *pharmakon*, or, that which is simultaneously destructive and curative (Stiegler, 2013, p. 4).
4. This is not to suggest that there was an identifiable rupture between the analog and digital eras. Andrea Bohlman and Peter McMurray remind of the overlaps between the two represented by Digital Audio Tape (DAT), as well as the usage of analog audio tape for data storage. Andrea F. Bohlman and Peter McMurray, "Tape: Or, Rewinding the Phonographic Regime," *Twentieth-Century Music* 14, no. 1 (2017).
5. According to Bruce Davis, the WSP had no explicit political agenda or intention of responding to the current cultural moment. Personal communication with Bruce Davis, August 2017.
6. Personal communications with Barry Truax and Bruce Davis, August 2017.
7. One WSP text contains a sort of appendix listing the "field recording equipment" used on a given project. However, it does not discuss the idea of field recording in any overt way. R. Murray Schafer, *Five Village Soundscapes* (Vancouver: A.R.C. Publications, 1977a), 331.
8. Randolph Jordan offers a nuanced reading of the WSP's approach to location recording from the early 1970s to the post-Schafer era of the late 1970s and beyond. Randolph Jordan, "Unsettling the Soundtrack: Acoustic Profiling and Documentation of Community and Place," in *The Routledge Companion to Screen Music and Sound*, ed. Miguel Mera, Ronald Sadoff, and Ben Winters (New York: Routledge, 2017). See also Jan Marontate, Megan Robertson, and Nathan Clarkson, "Soundscapes as Commemoration and Imagination of the Acoustic Past," in *Rouledge International Handbook of Memory Studies*, ed. Anna Lisa Tota and Trevor Hagen (New York: Routledge, 2015).

9. On the material substrate that makes digital technology possible, see Matthew Kirschenbaum, *Mechanisms: New Media and the Forensic Imagination* (Cambridge: MIT Press, 2008).
10. The obvious touchstone for this line of thought is Judith Butler, *Gender Trouble: Feminism and the Subversion of Identity* (New York: Routledge, 1999).
11. Bruno Latour, "Telling Friends from Foes in the Time of the Anthropocene," in *The Anthropocene and the Global Environment Crisis – Rethinking Modernity in a New Epoch*, ed. Clive Hamilton, Christophe Bonneuil, and François Gemenne (London: Routledge, 2015).

References

Anderson, B. (2006). *Imagined Communities: Reflections on the Origin and Spread of Nationalism*. London: Verso.

Berton, P. (1997). *1967: The Last Good Year*. Toronto, Canada: Doubleday.

Bohlman, A. F., & McMurray, P. (2017). Tape: Or, Rewinding the Phonographic Regime. *Twentieth-Century Music, 14*(1), 3–24.

Butler, J. (1999). *Gender Trouble: Feminism and the Subversion of Identity*. New York: Routledge.

Conner-Simons, A. (2016, June). Artificial Intelligence Produces Realistic Sounds That Fool Humans. *MIT News*. Retrieved from http://news.mit.edu/2016/artificial-intelligence-produces-realistic-sounds-0613

Davis, B., & Huse, P. (1974). Cross-Canada Soundscape Tour 1973. *Sound Heritage, III*(4), 28–48.

Jameson, F. (2005). *Archaeologies of the Future: The Desire Called Utopia and Other Science Fictions*. London: Verso.

Jordan, R. (2017). Unsettling the Soundtrack: Acoustic Profiling and Documentation of Community and Place. In M. Mera, R. Sadoff, & B. Winters (Eds.), *The Routledge Companion to Screen Music and Sound* (pp. 590–602). New York: Routledge.

Kirschenbaum, M. (2008). *Mechanisms: New Media and the Forensic Imagination*. Cambridge, MA: MIT Press.

Latour, B. (2015). Telling Friends from Foes in the Time of the Anthropocene. In C. Hamilton, C. Bonneuil, & F. Gemenne (Eds.), *The Anthropocene and the Global Environment Crisis – Rethinking Modernity in a New Epoch* (pp. 145–155). London: Routledge.

Macdonald, C. (2004). Ultra-Red. *Grooves Magazine*, p. 16.

Manovich, L. (2001). *The Language of New Media*. Cambridge, MA: MIT Press.

Marontate, J., Robertson, M., & Clarkson, N. (2015). Soundscapes as Commemoration and Imagination of the Acoustic Past. In A. L. Tota & T. Hagen (Eds.), *Rouledge International Handbook of Memory Studies* (pp. 519–532). New York: Routledge.

Massumi, B. (1987). Translator's Forward. In *A Thousand Plateaus: Capitalism and Schizophrenia* (pp. ix–xv). Minneapolis, MN: University of Minnesota Press.

Mumford, L. (1922). *The Story of Utopias*. New York: Boni and Liveright.

Plato. (1926). *Laws* (The Loeb Classical Library). Cambridge, MA: Harvard University Press.

Robinson, D. (2008). Analog. In M. Fuller (Ed.), *Software Studies: A Lexicon* (pp. 21–31). Cambridge, MA: MIT Press.

Schafer, R. M. (1967). *Ear Cleaning: Notes for an Experimental Music Course*. Toronto, Canada: Clark & Cruickshank.

Schafer, R. M. (1974). Soundmarks. On *Soundscapes of Canada*, Programme III [Radio Program, CD]. Vancouver, Canada.

Schafer, R. M. (1977a). *Five Village Soundscapes*. Vancouver, Canada: ARC Publications.

Schafer, R. M. (1977b). *The Tuning of the World*. New York, NY: Knopf. Reprinted as *The Soundscape: Our Sonic Environment and the Tuning of the World*. Rochester, VT: Destiny Books, 1994.

Schafer, R. M. (1984). *On Canadian Music*. Bancroft, Canada: Arcana Editions.

Spivak, G. C. (1996). *The Spivak Reader*. New York: Routledge.

Stiegler, B. (2013). *What Makes Life Worth Living: On Pharmacology*. Cambridge, MA: Polity.

Truax, B. (1978). *Handbook for Acoustic Ecology*. Vancouver, Canada: ARC Publications.

Truax, B. (1984). *Acoustic Communication*. Norwood, NJ: Ablex Publishing Corporation.

Truax, B. (1996). Soundscape, Acoustic Communication and Environmental Sound Composition. *Contemporary Music Review, 15*(1), 49–65.

Ultra-red. (2008a). *10 Preliminary Theses on Militant Sound Investigation*. New York: Printed Matter.

Ultra-red. (2008b). *Constitutive Utopias: Sound, Public Space and Urban Ambience*. http://www.temporaryservices.org/ultratext.html

Ultra-red. (2008c). *Some Theses on Militant Sound Investigation, or, Listening for a Change*. https://inthemiddleofthewhirlwind.wordpress.com/some-theses-on-militant-sound-investigationor-listening-for-a-change/

Westerkamp, H. (2002). Linking Soundscape Composition and Acoustic Ecology. *Organised Sound, VII*(1), 54.

CHAPTER 7

Havana's Falling Tanks and Flooded Laneways: Examining the Acoustic Community

Vincent Andrisani

Introduction

Aside from the term "soundscape", there are few concepts in the discipline of soundscape studies that have been both as useful and as contentious as "acoustic community". Developed through the pioneering work of the World Soundscape Project (WSP, 1973, 1977, 1978), acoustic community, a team of researchers at Simon Fraser University led by composer and educator R. Murray Schafer, refers, broadly, to social formations generated in sound and listening. It offers a general model for describing not just interpersonal forms of communication, but shared sonic experiences. "Community can be defined in many ways: as a political, geographical, religious, or social entity", observed Schafer, "but I…propose that the ideal community may also be defined advantageously along acoustic lines" (1977, p. 215). Conceived of as a "bounded system" defined on the basis of information exchange (Truax, 2001, 2009, 2017), the acoustic community is inclusive of all individuals in a given context—regardless of sentiment or allegiance—who engage in the practice of soundmaking, listening, or both. At the time it emerged, the concept proposed a previously unexplored

V. Andrisani (✉)
Simon Fraser University, Vancouver, BC, Canada

M. Droumeva, R. Jordan (eds.), *Sound, Media, Ecology*,
Palgrave Studies in Audio-Visual Culture,
https://doi.org/10.1007/978-3-030-16569-7_7

means not only of studying sound, but also of conceiving of community, and was born out of an interdisciplinarity that indeed challenged existing scholarly paradigms. And because the term "community" still holds currency in contemporary research, acoustic community has tremendous potential as a scholarly concept even today.

However, it raises just as many questions as it provides answers. This ambiguity is not specific to the concept of the acoustic community per se, but is true of almost any scholarly use of the term "community". "Community", sociologist Tony Blackshaw (2010) observes, is "one of the most vague and imprecisely drawn concepts in the social sciences; it seems to mean everything and nothing" (p. 2). As both a colloquial term and a scholarly concept, community necessarily refers to a social bond that brings people together, but there is no consensus about the precise nature of that bond. Is geographical proximity sufficient? Must it involve some form of political solidarity? Or is it, as Benedict Anderson (1983) argues, a symbolic allegiance that is immaterial and "imagined"? These questions have fuelled continued debates over the merits of "community", how it can be used, and whether it should be used at all. Of course, the WSP had their own interpretation of what brings people together, the spirit of which emerged out of the group's broader research trajectory and was influenced by the era's scholarly debates—two attributes that are not necessarily shared by all sound studies researchers. This became particularly evident to me while developing my own research, which is based in (and on) Havana, Cuba, a city firmly grounded in the Latin American and Caribbean region.

As a sonic ethnographer trained in the tradition of soundscape studies, I am interested in the ways that the concept of the acoustic community can be put to use in present-day, qualitative research. But rather than simply transplant a theory generated in the developed world to the Caribbean, I instead want to do the reverse: to ask what it means to theorize *from* the Caribbean by letting my ethnographic observations resonate within soundscape theory.[1] How can the concept of the acoustic community help us understand cities in Latin America and the Caribbean? And how do cities in Latin America and the Caribbean encourage us to revisit, revise, and expand what we already know about acoustic communities? I begin by historicizing the concept, which entails unpacking the work of the WSP and the idea of modernity to which it prescribes. I then contrast this historiography with the conflicting conceptions of Latin American modernity, which lead to very different interpretations of the term community and, by extension, acoustic community. I end by elucidating a notable example of an acoustic community that emerges time and again in Havana linked to the issues of overpopulation and infrastructural degradation. By intersecting soundscape studies with my ethnographic experiences, I ultimately argue for the use of "acoustic community" in a manner that is contingent upon the historical geography on which it is used.

The Origins of the Acoustic Community

In order to work with the concept of the acoustic community, we must first account for the origins of the term and the research trajectory from which it emerged. This entails exploring the work of the World Soundscape Project, which began in Vancouver in the 1960s and early 1970s. Through a pioneering study titled *The Vancouver Soundscape* (1973), the group gave life to some of the terms, concepts, and methodological techniques that would eventually come to define soundscape studies. For instance, "signals", "soundmarks", and "keynote sounds" emerged through the exploration of the city's sonic geographies. Putting these concepts to use, the group observed that "signals and soundmarks define the acoustic spaces of the community" (p. 41), referring to the capacity for sound to articulate the social life of a given area or neighbourhood. However, increasing levels of automobile traffic, seaplanes, and building exhaust masked these discrete sounds, making them difficult to discern. The result, the group surmised, was that the soundscape was becoming "more monotonous" (p. 33), and the potential for "community" to emerge in sound and listening was imperilled. To further interrogate these observations, the WSP decided to examine geographies of a much smaller scale using settlements that are both less populated and characterized by fewer broadband noises.

This brought the group to Europe, where they explored the soundscapes of small villages in Italy, Sweden, France, Scotland, and Germany. What emerged was *Five Village Soundscapes* (1977), a series of case studies through which they believed the historical trajectory of much larger settlements could be theorized. The five selected villages varied in size, population, urban design, and economic development, to say nothing of the diverse cultural practices of inhabitants themselves. Using interviews with residents (or "earwitness accounts"), field recordings, soundwalking, and simply listening, the group developed reflections on the rhythms and patterns of everyday life, the acoustic definition of village spaces, and the community attitudes towards the soundscape. In so doing, they paid particular attention to the extent to which each village was subject to external economic, social, and cultural pressures. "Change in acoustic definition of a community runs parallel to social change", the group observed, "and in our experience, the change in western [sic] industrial society is always toward reduced definition" (1977, p. 49). The resulting impression, in terms of the acoustic community, was that the influx of industrial technologies and machine-generated sound is inversely related to the health and viability of community life.

The WSP's case studies, which consist of *The Vancouver Soundscape* (1973) and the subsequent study of *Five Village Soundscapes* (1977), led to the development of a broader historiographic narrative around which the group's research gravitated. This narrative is articulated with the greatest amount of depth in Schafer's seminal text, *The Tuning of the World* (1977), republished in 1994 as *The Soundscape: Our Environment and the Tuning of the World*. Animated by the historical moment of industrialization, Schafer's history posits the existence of pre- and post-industrial soundscapes and warns of the potential consequences of machine-generated noise on the acoustic environment. The approach takes cues from the modern environmental movement which, for the first time in the 1960s and 1970s, galvanized public concern for environmental degradation in places like the United States, Japan, and other parts of the globe. The result is that, like the natural ecologist, the acoustic ecologist (or soundscape researcher) advocates for soundscapes that function on a "human scale" and are characterized by "variety, complexity, and balance"—traits that are characteristic of both a healthy ecosystem and a functional acoustic community alike. So, for Schafer and his team, and for most acoustic ecologists today, acoustic communities emerge as moments of acoustic clarity, audibility, and definition, and are evaluated against the potentiality of machine-generated noise.

That the work of the WSP gravitates so heavily around industrial modernity has raised concerns for a number of scholars who have called attention to Schafer's "environmental romanticism" (Kunreuther, 2017) and its articulation across the theory and discourse of soundscape studies (Porcello, Meintjes, Ochoa, & Samuels, 2010; Thompson, 2002). This includes the concept of the acoustic community, which invokes an anti-industrial, environmental politics as it existed in the 1970s. Such a perspective is by no means unique to the WSP, but is in step with other scholarly treatments of "community" developed during the same era. In sociology, for instance, as Talja Blokland (2017) points out, there was a strong tradition throughout much of the twentieth century that posited the historical demise of community alongside the process of urbanization. "As cities expanded with tremendous speed and became the icons of vice, disintegration, anonymity and immorality", observes Blokland, "the fear that communities would disappear was an immediate result and was seen from the outset as connected with urban development" (p. 16). The work of the WSP extends this tradition into the world of sound by likening pastoral geographies to functional community spaces, while an urban lifestyle tends to take communities out of balance. It is this tension between pre-modern and modern, rural and urban, clarity and unintelligibility, that prompts urban studies scholar Sophie Arkette (2004) to describe Schafer's work not as "romantic" per se, but, instead, as having an "urban prejudice".

Because the WSP conceives of community as a social formation under threat, the concept of the acoustic community tends to emphasize form over content. That is, community formation depends mainly upon the shared experience of discernible sounds, no matter what those sounds might be. For the WSP, the ability to attend to and interpret discrete acoustic signals is itself an enactment of resilience in an increasingly industrializing and urbanizing world. However, this approach remains grounded in a mid-twentieth-century reading of Western modernity, and is troubled by histories and geographies in which industrialization is not the modern hegemonic project. In such cases, community is dependent not upon a bond generated by interpersonal communication, but on political ties that emerge as local responses to regional and global enactments of power. To acknowledge the local dynamics of political struggle is the mandate of any qualitative researcher, and is manifest in both the scholarship they develop and in the theoretical tools they use. So, as someone investigating the intersection of sound, listening, and community in the city of Havana, my research raises the following questions: how does Latin American modernity, which does not gravitate around the consequences of industrialization, trouble the conception of the acoustic community? How can we conceive of community when it involves much more than intelligibility and information exchange alone? And in light of these concerns, how then do we define community?

Latin American Modernity, Urban Informality, and the Making of Community

Conceptions of Western modernity have historically not been troubled by the inescapable global relationality that conditions theories of the Global South. This ambivalence is noteworthy, as scholars have repeatedly shown (Mintz, 1986; Trouillot, 1995; Williams, 1961), on account of how historically pivotal the Global South was (and remains still today) in the rise of Western dominance. This is particularly true of Latin America and the Caribbean which, through the exploitation of labour and resources, provided one of the main sources of European economic accumulation at the onset of early capitalism. In spite of how integral these geographies are to the history of Western modernity, conceptions of the historical epoch tend to characterize it as a moment of industrialization and urbanization alone. By contrast, Latin American and Caribbean modernity, although inclusive of industrialization and urbanization, consists primarily of European conquest and American expansionism. It is defined by the enactment of colonial power, the influence of which extends right up to the present day. The result is that prevailing conceptions of Latin American and Caribbean modernity describe it as an ongoing historical moment that is "contradictory" and "hybrid" (García Canclini, 1995), possessing a "dark side" (Mignolo, 2002) and "an articulation of many rationalities" (Quijano, 2000).

The unequal conditions of Latin American and Caribbean modernity are made most tangible in urban centres across the region. Overpopulation, the growth of slum areas, a crisis in housing, infrastructural ruination, and the scarcity of basic necessities are only some of the struggles that residents face. The result is a form of urbanism that is itself, as García Canclini describes, "contradictory" and "hybrid". On the one hand, there is the formal city which, as Fischer, McCann, and Auyero (2014) observe, "profits economically and politically from the slum's illegality" (p. 1). This city is governed by policy at both the municipal and state levels and is conditioned by a market to which most residents cannot gain uninterrupted access. On the other hand, there is the informal city, which "provides a refuge from utopian urban regulations that would otherwise exclude many poor Latin Americans from city life" (p. 1). Referring less to a particular spatial geography than to the ways that residents creatively resolve their most pressing needs, the informal city is animated by the collective energies of the city's marginal communities. Such practices are referred to as "informal urbanism", which, urban studies scholar Rahul Mehrotra (2010) observes, is about "invention within strong constraints with indigenous resources with the purpose of turning odds into a survival strategy" (p. xiii).

Although informality is an important part of everyday life in cities across the region, it takes on a special quality in the city of Havana. Here, residents are known for their creative responses to the island's severely limited access to international capital on account of the U.S. embargo. Remittances from family and friends living abroad, alongside the slowly increasing potential for entrepreneurial activity, help support financial flows; however, they are generally insufficient, making the need to engage informal practices necessary for most residents. Such practices are described colloquially using the term *inventar*: to invent on account of the city's tenuous social, economic, and material conditions. *Inventar* is most commonly demonstrated through ingenuity with material objects, which could entail giving new life to out-of-date and otherwise defective consumer technologies (1950s' American automobiles being the most famous example). Or, it could entail responding to the city's housing crisis by creating makeshift spaces in which to live (del Real & Pertierra, 2008). In its most enterprising form, it is manifest as a street-based, money-making venture through either sex work or street hustling (Carter, 2008). But in every case, *inventar* requires not only ingenuity, spontaneity, and wit, but also a broader social network in order to be realized.

It is against the backdrop of informality, or *inventar* in the case of Havana, that community formation in cities across Latin America and the Caribbean must be evaluated. No matter how commonplace and seemingly inconsequential, everyday social interactions are often engaged with the aim of mitigating economic and material precarity. "Community matters more in low-income areas", argues urban sociologist Anne Power (2007), "because families have less cash to buy support, need local back-up in emergencies and place a high value on local social contact because wider networks beyond the immediate community are harder to reach and maintain on a low income" (p. 45). In Havana, everyday dialogue often gravitates around gaining access to material goods and/or services. For instance, individuals with access to supplies of food items make their rounds through neighbourhoods to subtly inquire about the public desire (and ability) to purchase them. Mobile vendors walk the streets hawking items such as mops, clothes hangers, and other housewares, which are particularly useful when the local market is sold out. And friends and neighbours who have the skills to work with material technologies can help repair household items such as a malfunctioning appliance or a plumbing fixture. Each of these opportunities must be located not in a store and certainly not through an internet search, but rather, during otherwise inconsequential, everyday social interactions. In this sense, participating in local communities is not just strategic; it is necessary in order to *sobrevivir*—to survive.

Conceptions of informality have traditionally been understood as a consequence of top-down articulations of power that undermine the political agency of a local population. In Havana, this perspective has given life to readings of the city as "dystopian" (Coyula, 2011), a "non-city" (Redruello, 2011), or as altogether "illegible" (Rojas, 2011). However, this tends to overlook the creativity, ingenuity, but above all, the potentiality of informal urban life to challenge the hegemony of the city. And residents do so collectively, not through protest or organized dissent, but, instead, through everyday, lived practices. In so doing, residents assert what urban theorist Saskia Sassen (2008) refers to as "presence" amidst the urban geography. "That the disadvantaged in global cities can gain 'presence' in their engagement with power but also vis-á-vis each other", Sassen observes, "does not necessarily bring power, but neither can it be flattened into some generic powerlessness" (p. 317). In this sense, the communities generated on the basis of urban informality cannot be conceived of as social formations in peril, in spite of the conditions of coloniality out of which they emerge. Instead, they are generative, and are in fact a result of Latin America and the Caribbean's unequal modernity. And as I argue, such communities are formed, in part, through communicatory practices in which sound and listening figure centrally.

Acoustic Communities in Havana

The study of sound in Havana brings together two separate yet interrelated research trajectories. On the one hand, there is sound-based urban research, which has given life to accounts of the city as a musical geography (Lashua, Spracklen, & Wagg, 2014; Peterson, 2012), in terms of its relationship to audio technology (Barns, 2014; Bull, 2000), and has offered new and previously unwritten histories through sound and listening (Mattern 2017; Valliant, 2003). On the other, there are sound studies of Latin America and the Caribbean, which have similarly developed accounts of the region's musical geographies (Madrid, 2011; Solomon,

2000), the role of the radio in the public sphere (Casillas, 2014; González, 2002), and have (re)thought regional histories on sonic terms (Bronfman, 2016; Bronfman & Wood 2012). The work of ethnomusicologist Ana María Ochoa Gautier (2012) is particularly important in this regard, which elaborates sound as a crucial realm through which Latin American modernity has been made and is being contested. "In speaking about Latin America as an aural region, I argue that under the contemporary processes of social globalization and regionalization coupled with the transformations in the technologies of sound, the public sphere is increasingly mediated by the aural" (2012, p. 392). The "aural public sphere", as Ochoa Gautier terms it, denotes sound-based forms of political participation that emerge in the context of everyday life. It is manifest in public discourse, in both mediated and unmediated forms of communication alike.

As a conceptual tool, the aural public sphere succinctly describes the politics of urban informality. It communicates the ways that power, resilience, and political agency are articulated in sound and listening. Yet, it does so in a way that deals less with the practice of informality in and of itself, and more with the symbolic potential of such practices. It refers not to the experience of resilience as much as it describes the resulting politics of that experience. For these reasons, urban informality and the sound-based modes of communication through which it emerges can perhaps more appropriately be described as acoustic communities. Community, in this sense, is defined not simply as a social formation that people belong to, but also as *something people do*. It is a cultural practice of affiliation and solidarity, and it emerges in the form of interpersonal, communicatory exchanges. "If community is culture, it needs to be understood as a relational figuration in constant change and movement... community means culture through communication" (Blokland, 2017, p. 45). The idea of community as cultural practice has gained traction in the fields of urban studies and cultural geography, yet the challenge of how to grasp the ephemeral nature of community still remains. This, I argue, is precisely the merit of sound and listening. More specifically, this is the merit of the acoustic community, which affords a means of both locating and unpacking the moments during which communities are made tangible.

Exploring the contours of the acoustic community begins with the act of listening. In Havana, to participate in acoustic communities demands a place-based form of listening that both acknowledges and accommodates the design of the built environment. Because it is hot in the Caribbean year-round, doors and windows that are not in direct sunlight are often left open. Additionally, windows are covered not with panes of glass, but with wooden shutters, which allow sound to permeate at any time of the day. The result is an urban geography that mediates, rather than inhibits, sound-based modes of communication. Sound, as Barry Blesser and Linda Ruth-Salter (2007) might put it, "illuminates" the spaces of the neighbourhood. Through windows, doors, from balconies, and through porticoes, residents engage in participatory exchanges that are often quite animated and lively. They do so without the necessity of corresponding visual cues, which demands an intimate knowledge of the built environment, the actors, and the social practices that bring it to life. I refer to this form of communication as *trans-liminal listening*, or, *listening trans-liminally* (which amount to the same thing): a mode of aural perception that extends beyond the threshold of the built environment. Listening in this way is integral to the formation of acoustic communities, and as I realized rather quickly during my time there, it is a learnt skill that takes time and experience to acquire.

While in Havana, I was captivated by the extent to which residents communicate through the open spaces of the built environment. People living on the second, third, and fourth floors of homes or apartments engage in casual conversation with neighbours whose windows or balconies are within speaking distance. Similarly, conversation between the street and the home is commonplace. From outdoors, people announce their presence to friends, family, or neighbours inside their home, and, conversely, from indoors, residents holler into the streets. These types of exchanges animate the vibrant social life of neighbourhoods across the city, as residents engage in what seem to be nothing more than everyday, rather ordinary forms of communication. Yet, more often than not, there is more to these moments than initially meets the ear. Upon further listening, questioning, and learning about everyday life in Havana, it became apparent that many trans-liminal communicatory exchanges functioned as tactical negotiations—as ways through which residents engage *inventar*. They are part of the everyday struggle (known colloquially as *la lucha*) and help mitigate the precarity that characterizes urban life. So, I began to wonder how sound and listening might articulate the negotiation of Havana's most pressing issues. For instance, how might everyday modes of communication be used to mitigate material scarcity? Overpopulation? Or architectural and infrastructural ruination?

Each of these issues conditions everyday life in the city, and each one emerges from a regional history that gravitates not around industrial modernity but, instead, around a history of colonial and imperial rule. This is the hegemonic project in Latin America and the Caribbean, and these are the political struggles around which local communities emerge. So, although the question of discernibility is indeed an important part of sound-based modes of communication, as the WSP rightly observed, it is not the central tenet of community life in Havana. As a sonic ethnographer exploring the relationship between sound, listening, and the ongoing conditions of coloniality, I am interested in understanding how and for what reason acoustic communities in Havana emerge. To what extent are they a response to the region's prevailing urban issues? And how might these sound-based communities constitute a new and unexplored dimension of acoustic community as a scholarly concept? In the following ethnographic exploration, I illustrate a particular moment that I encountered time and again in my neighbourhood in the district of El Vedado. The moment crystallizes the issue of architectural and infrastructural degradation, the practice of listening trans-liminally, and the unequal modernity characteristic of life not only in Havana, but in cities across the region. The result is an acoustic community born of resilience that asserts, what Sassen (2008) might term, the embodied "presence" of residents in the spaces of the city.

Falling Tanks and Flooded Laneways

While in Havana, I woke up almost every day at the same time and in the same way. At about 6:45 am, a slow, steady waterdrop began falling onto the air conditioner in my bedroom window. The shape and density of the air conditioner's aluminium enclosure, combined with the height from which the water was falling, created a loud, thud-like sound that made it impossible to sleep. *Bloop. Bloop. Bloop.* If I listened long enough, I began to hear the sound in new ways. *Smack!* Then, *womp womp womp.* The enclosure resonates after the water lands. Initially, I thought these were the sounds of rain. But without fail, I'd walk out onto the balcony and see nothing in the sky but the hot Caribbean sun. So I asked my roommate about it, and he promptly led me up the stairs of the apartment and out onto the rooftop. There, sat a large communal reservoir that supplies water to each apartment in the building. It's filled twice daily, and each time, almost without fail, it overflows. The excess water makes its way across the rooftop and over the side of the building. Along its downward route is my window, where it slowly accumulates, causing the early-morning dripping. Through both drainage pipes and alongside the outer wall of the building, the water then makes its way into the laneway below, which it then floods before it falls into the sewer, completing its journey through the neighbourhood.

Directly affected by the errant streams and flows, however, are the residents who live at ground level in the building next door. Once a single-family, fully detached home, the building has since been converted into an apartment that now houses upward of twenty or twenty-five people. By Havana's standards, it's not very old, but the numerous modifications it's undergone over the years have put added stress on the structure—to the point that it now has a series of wooden planks leaning against the outer wall for additional support. One such modification is that the laneway is now the entrance to a number of small makeshift apartments known as *cuarterias*. When the laneway is flooded, it becomes inconvenient, if not altogether unmanageable, for residents to enter and exit without stepping in the water. For a short period of time, an important part of their already limited living space is unusable. To make matters worse, these residents often hang their laundry to dry in the precise location that water tends to fall. More than simply an occasional, inconsequential episode, the water runoff is a daily disruption that affects a number of households in the community.

For residents, this sequence of events issues a not-so-subtle reminder of the precarity of the built environment. On the one hand, the overflowing occurs as a result of a water supply system that is less than fully functional. And on the other hand, residents most immediately affected by the overflow live in makeshift apartments that are both structurally deteriorated and situated in undesirable locations. The water overflow and runoff are therefore not just representative of a haphazard sequence of events within the confines of this neighbourhood alone. Rather, this seemingly insular encounter is symptomatic of a series of broader urban and regional issues that include overpopulation and infrastructural and architectural ruination. Taken together, these conditions comprise an "urban crisis" (Clark, 1998) that characterizes not only the city of Havana, but cities across Latin America and the Caribbean.

The conditions of ruination and overpopulation in my neighbourhood compel residents to find ways of mitigating the daily overflow and runoff. One way they do so is in sound and listening, which is precisely how I became aware of this situation in the first place. In particular, residents mobilize acoustic communities around the overflow and runoff as a way of mitigating the disruption. It unfolds as follows.

Filling the reservoir usually takes place at 5 or 5:30 am, and then again at 2 or 2:30 pm. It is looked after by our neighbour Ariel whose apartment is on the fifth and top floor, nearest the reservoir. Using a switch from inside his home, Ariel activates a motorized pump in the parking garage below (as seen in Fig. 7.1), which draws water up through the supply lines into the tank above. Residents gain access to the water by simply turning on one of the faucets in their home or by flushing the toilet, allowing gravity to take its course. Typically, the reservoir takes a couple of hours to fill, depending on the amount of water used over the course of the day. Once the reservoir is full, however, it overflows quickly and without warning.[2] The water then travels across the rooftop and over the side of the building where it traverses a number of cables, windows (including my own), and pipes before it floods the laneway below. Ariel does his best to prevent the spilling by turning off the pump before the tank overflows, but even on a good day, a small amount finds its way onto this expected route.

Fig. 7.1 Motorized water pump at ground level in my residence. (Photo by V. Andrisani, (author))

Without fail, this series of events brings one or even several residents who live in the makeshift apartments out of their homes and into the laneway. They immediately shout up to Ariel, aiming to notify him—and everyone else in the community—of yet another overflow. One resident in particular predictably whistles three times before yelling the phrase *"¡se bota el tanque!"*, which translates literally to "the tank is falling" but means "the water is overflowing!" At times, neighbours who live in second-, third-, and fourth-floor apartments will chime in. These residents either repeat the call of those in the laneway, expediting the message up to Ariel, or they express their own discontent with yet another disturbance by saying things like *"¡oye llevo dos horas avisandole!"*, meaning "I've been saying [the water is overflowing] for the last two hours!" This of course is an exaggeration because the disruption begins and ends within a matter of about fifteen minutes. But it illustrates residents' frustration with the daily occurrence, which is expressed by community members who are not even directly affected by it. Ariel of course acknowledges the calls, first by responding vocally, then by turning off the motor.[3]

Clearly, sound plays an integral role in this sequence, if for no other reason than by way of the dialogue itself. By shouting into and out of the home, residents communicate with one another in a timely manner. Doing so allows those at ground level to immediately alert community members of another water mishap. So, by hollering up to Ariel from the laneway, ground-level residents declare something to the effect of "the water is overflowing, turn off the pump". But in another sense, the act of shouting up to Ariel—and to the rest of the community—implies something more akin to "these spaces are being flooded, not only do I live here, but *I am here*". It represents not only a means through which to redraw the limits of the neighbourhood's living spaces and recover them in a timely manner, but it is also a symbolic gesture enacted with the intent of reasserting an embodied presence in one's own living space.

Yet residents listen to sounds that extend beyond those of vocal communication alone. Before the shouting begins, and even before the water makes its way into the laneway, residents are already aware of the activity taking place within their neighbourhood. The first and most obvious sound associated with filling the reservoir is the broadband, steady-state drone of the water pump that sits at ground level. Generated by an electric motor, this sound is quite loud, and it announces to most residents in the neighbourhood that the reservoir is being filled. Once full, the tank's overflow begins to splatter on the rooftop (see Fig. 7.2), but only those

Fig. 7.2 Water overflowing from the spout of the rooftop reservoir. (Photo by V. Andrisani, (author))

who live nearby can hear it. Following its descent, however, the water bubbles and splashes as it springs from drainage pipes and flows over the cement floor of the laneway. For ground-level residents, these are the most urgent sounds. They indicate that the laneway is being flooded and are the final acoustic call to action. Eventually, the flowing water empties into the sewer, but by the time residents hear it, the water has already completed its route through the spaces of the neighbourhood (see Fig. 7.3).

The numerous sounds produced over the course of this sequence enable residents to accurately follow the water along its expected route, even if they are not physically co-present with it. Through open windows and doors, residents listen trans-liminally to the sounds of supply

Fig. 7.3 The flooded laneway. (Photo by V. Andrisani, (author))

infrastructure and remain attuned to the possibility of laneway flooding. At the outset, the motorized water pump makes residents aware that the reservoir is being filled. So, by the time the water arrives in the laneway, usually about an hour and a half later, residents are prepared. This immediately brings those in ground-level *cuarterias* outside, who then notify the rest of the neighbourhood. The sooner ground-level residents hear the water, the sooner everyone alerts Ariel, and, ultimately, the sooner the flooding comes to an end. In so doing, residents take the opportunity to partially—and at times, almost entirely—mitigate the flooding. Long before the shouting begins, the act of listening is a means through which residents assert their embodied presence in the neighbourhood. It is a tacit first step in a series of preventative measures that safeguard their living spaces. And it is accomplished through a form of everyday listening that remains attuned to the sounds of the neighbourhood, articulated in the form of an acoustic community.

Conclusion

The concept of acoustic community plays an important role in my research on the city of Havana. It offers a succinct and fertile means through which to describe the collective practices that bring to bear the lifeworld of the neighbourhood. It also provides the basis for theorizing the modes of listening that are integral to the social life of the city which, in Havana, consist of *listening trans-liminally*: listening across the spatial threshold of the built environment. But any use of the term "community" in Latin America and the Caribbean must necessarily acknowledge the colonial and imperial history that comprises the backdrop of the region's everyday life today. It must acknowledge that modernity itself is an unequal global-historical epoch. For this reason, I conceive of Havana's acoustic communities not on the basis of acoustic discernibility alone but also in terms of the shared struggle that residents engage in order to mitigate the precarious conditions of everyday life in the city. As a result, community, as I've developed the term here, gravitates around informal practices, or *inventar* in the case of Havana. The episode of water overflow and runoff in my neighbourhood simultaneously makes tangible the precarious conditions of the region's modernity, and the ways that residents creatively and collectively negotiate these issues. Such practices offer an added dimension to the term acoustic community, and to accommodate them demands an expanded conception of the term.

In spite of the unavoidable imprecision and ambiguity of "community", it still remains an often-used and highly useful conceptual tool in scholarship across the human sciences. Scholarly work surrounding the contours of community formation continues to emerge along myriad scholarly trajectories and in most any possible global-historical context. The central struggle of many of these studies is the ability to discern community as a social formation that is always in flux, always changing, never static. I maintain that a sound-based approach has the capacity to offer a perspective that captures community formation in motion, as an ephemeral and fleeting process. For this reason, the acoustic community is a theoretical tool that has much to contribute not only to the study of sound and listening but also to community studies more broadly. However, in order to do so, the term must be situated according to the historical geography(ies) in which it is used. The idea of community must acknowledge and accommodate the hegemonic project(s) in which local populations are embedded and against which collective political allegiances emerge. This is precisely what I aimed to do in this chapter, and in my work in Havana more broadly, yet there remains much to be learned about acoustic communities and how the term may be utilized in contexts and for populations yet to be considered.

Notes

1. Here, I take inspiration from the work of urban studies scholars Ananya Roy (2005, 2009) and Jennifer Robinson (2014), which offers a reflexive counterpoint to Euro-American theoretical elaborations by conceiving of cities in the Global South as the "new geographies of theory".
2. The reason that it continually overflows is because the float ball assembly that regulates filling is broken. Ironically, this assembly and the motorized system is relatively new in comparison to what is used throughout the rest of the city. Presumably, because it is not widely used, finding the parts to repair the fill valve has been challenging. As such, residents have simply learnt to make do with the faulty system until a new float ball assembly becomes available.
3. This type of hyperbole is not uncommon in Cuba. It's a popular form of dialogue known as *choteo*: informal humour that explicitly targets authority with the aim of undermining it. About *choteo*, scholar Damián J. Fernández (2000) writes, "The *choteo* deauthorizes authority by debunking it and constitutes a form of rebellion. It is undisciplined, unserious, even if the business at hand is of the utmost importance. It reflects contempt for and cynicism about higher-ups and the institutions of society...the purpose is to use humour as a way to privatize social relations by making them accessible, at least momentarily, by bringing the people and the institutions that stand above the common folk down to the level of the popular, of the streets, of "us". This is *choteo's* equalizing effect" (p. 31).

References

Anderson, B. R. O. (1983). *Imagined Communities: Reflections on the Origin and Spread of Nationalism*. London: Verso.

Arkette, S. (2004). Sounds Like City. *Theory, Culture and Society, 21*(1), 159–168.

Barns, S. (2014). Sounds Different: Listening to the Proliferating Spaces of Technological Modernity in the City. *Space and Culture, 17*(1), 4–15.

Blackshaw, T. (2010). *Key Concepts in Community Studies*. Los Angeles/London: Sage.

Blesser, B., & Salter, L.-R. (2007). Spaces Speak, Are You Listening? Experiencing Aural Architecture. Cambridge, MA: MIT Press.

Blokland-Potters, T. (2017). *Community as Urban Practice*. Malden, MA: Polity Press.

Bronfman, A. (2016). *Isles of Noise: Sonic Media in the Caribbean*. Chapel Hill, NC: The University of North Carolina Press.

Bronfman, A., & Wood, A. G. (Eds.). (2012). *Media, Sound, & Culture in Latin America and the Caribbean*. Pittsburgh, PA: University of Pittsburgh Press.

Bull, M. (2000). *Sounding Out the City: Personal Stereos and the Management of Everyday Life*. Oxford/New York: Berg.

Carter, T. F. (2008). Of Spectacular Phantasmal Desire: Tourism and the Cuban State's Complicity in the Commodification of Its Citizens. *Leisure Studies, 27*(3), 241–257.

Casillas, D. I. (2014). *Sounds of Belonging: U.S. Spanish-Language Radio and Public Advocacy*. New York: New York University Press.

Clark, J. (1998). *The Housing Dimension in Cuba's Urban Crisis: Havana as a Case in Point*. Retrieved from http://housingcuba.blogspot.ca/p/cubas-urban-crisis_23.html

Coyula, M. (2011). The Bitter Trinquennium and the Dystopian City: Autopsy of a Utopia. In A. Birkenmaier & E. K. Whitfield (Eds.), *Havana Beyond the Ruins: Cultural Mappings After 1989* (pp. 31–52). Durham, NC: Duke University Press.

del Real, P., & Pertierra, A. C. (2008). Inventar: Recent Struggles and Inventions in Housing in Two Cuban Cities. *Buildings & Landscapes: Journal of the Vernacular Architecture Forum, 15*, 78–92.

Fernández, D. J. (2000). *Cuba and the Politics of Passion*. Austin, TX: University of Texas Press.

Fischer, B., McCann, B., & Auyero, J. (Eds.). (2014). *Cities From Scratch: Poverty and Informality in Urban Latin America*. Durham, NC/London: Duke University Press.

García Canclini, N. (1995). *Hybrid Cultures: Strategies for Entering and Leaving Modernity*. Minneapolis, MN: University of Minnesota Press.

González, R. (2002). *Llorar es un Placer*. La Habana, Cuba: Editorial Letras Cubanas.

Kunreuther, L. (2017). Democratic Soundscapes. *The Avery Review*, (21), 17–27.

Lashua, B., Spracklen, K., & Wagg, S. (2014). *Sounds and the City: Popular Music, Place, and Globalization*. Houndmills, Basingstoke/Hampshire, UK: Palgrave Macmillan.

Madrid, A. L. (Ed.). (2011). *Transnational Encounters: Music and Performance at the U.S.-Mexico Border*. New York: Oxford University Press.

Mattern, S. C. (2017). *Code + Clay...Data + Dirt: Five Thousand Years of Urban Media*. Minneapolis, MN/London: University of Minnesota Press.

Mehrotra, R. (2010). Foreword. In *Rethinking the Informal City: Critical Perspectives from Latin America* (pp. xi–xiv). New York/Oxford: Berghahn Books.

Mignolo, W. D. (2002). The Geopolitics of Knowledge and the Colonial Difference. *South Atlantic Quarterly, 101*(1), 57–96.

Mintz, S. W. (1986). *Sweetness and Power: The Place of Sugar in Modern History.* New York: Penguin Books.

Ochoa Gautier, A. M. (2012). Social Transculturation, Epistemologies of Purification and the Aural Public Sphere in Latin America. In J. Sterne (Ed.), *The Sound Studies Reader* (pp. 388–404). London/New York: Routledge.

Peterson, M. (2012). *Sound, Space, and the City: Civic Performance in Downtown Los Angeles.* Philadelphia: University of Pennsylvania Press.

Porcello, T., Meintjes, L., Ochoa, A. M., & Samuels, D. W. (2010). The Reorganization of the Sensory World. *Annual Review of Anthropology, 39*(1), 51–66.

Power, A. (2007). *City Survivors: Bringing up Children in Disadvantaged Neighbourhoods.* Bristol, UK: Policy Press.

Quijano, A. (2000). Coloniality of Power and Eurocentrism in Latin America. *International Sociology, 15*(2), 215–232.

Redruello, L. (2011). Touring Havana in the Work of Ronaldo Menéndez. In A. Birkenmaier & E. K. Whitfield (Eds.), *Havana Beyond the Ruins: Cultural Mappings After 1989* (pp. 229–245). Durham, NC: Duke University Press.

Robinson, J. (2014). New Geographies of Theorizing the Urban: Putting Comparison to Work for Global Urban Studies. In S. Parnell & S. Oldfield (Eds.), *The Routledge Handbook on Cities of the Global South* (pp. 57–70). New York: Routledge.

Rojas, R. (2011). The Illegible City: Havana After the Messiah. In A. Birkenmaier & E. K. Whitfield (Eds.), *Havana Beyond the Ruins: Cultural Mappings After 1989* (pp. 119–134). Durham, NC: Duke University Press.

Roy, A. (2005). Urban Informality: Toward an Epistemology of Planning. *Journal of the American Planning Association, 71*(2), 147–158.

Roy, A. (2009). The 21st-Century Metropolis: New Geographies of Theory. *Regional Studies, 43*(6), 819–830.

Sassen, S. (2008). *Territory, Authority, Rights: From Medieval to Global Assemblages.* Princeton, NJ: Princeton University Press.

Schafer, R. M. (1977). *The Tuning of the World.* Toronto, Canada: McClelland and Stewart. Reprinted as *The Soundscape: Our Sonic Environment and the Tuning of the World.* Rochester, VT: Destiny Books, 1994.

Solomon, T. (2000). Dueling Landscapes: Singing Places and Identities in Highland Bolivia. *Ethnomusicology, 44*(2), 257.

Thompson, E. A. (2002). *The Soundscape of Modernity: Architectural Acoustics and the Culture of Listening in America, 1900–1933.* Cambridge, MA: MIT Press.

Trouillot, M.-R. (1995). *Silencing the Past: Power and the Production of History*. Boston, MA: Beacon Press.

Truax, B. (2001). *Acoustic Communication* (2nd ed.). Westport, CT: Ablex.

Truax, B. (2009). Introduction to the Five Village Soundscapes (1977). In H. Järviluoma, M. Kytö, B. Truax, H. Uimonen, N. Vikman, & R. M. Schafer (Eds.), *Acoustic Environments in Change/Five Village Soundscapes* (2nd Rev. ed., pp. 286–289). Tampere, Finland: TAMK University of Applied Sciences.

Truax, B. (2017). Acoustic Space, Community, and Virtual Soundscapes. In M. Cobussen, V. Meelberg, & B. Truax (Eds.), *The Routledge Companion to Sounding Art* (pp. 253–263). New York: Routledge.

Vaillant, D. (2003). Peddling Noise: Contesting the Civic Soundscape of Chicago, 1890–1913. *Journal of the Illinois State Historical Society (1998-)*, *96*(3), 257–287.

Williams, E. E. (1961). *Capitalism & Slavery*. New York: Russell & Russell.

World Soundscape Project. (1973). *The Vancouver Soundscape* (R. M. Schafer, Ed.). Burnaby, Canada: World Soundscape Project, Sonic Research Studio, Department of Communication Studies, Simon Fraser University.

World Soundscape Project. (1977). *Five Village Soundscapes* (R. M. Schafer, Ed.). Vancouver/Burnaby, Canada: ARC Publications.

World Soundscape Project. (1978). In B. Truax (Ed.), *The World Soundscape Project's Handbook for Acoustic Ecology* (1st ed.). Vancouver, Canada: ARC Publications.

CHAPTER 8

Acoustic Ecology and Ecological Sound Art: Listening to Changing Ecosystems

Leah Barclay

The impacts of climate change are impossible to ignore. Devastating coral bleaching, ocean acidification and melting polar ice sheets continue to permeate media headlines. Forest degradation, air pollution, heat waves and species extinction are causing dramatic ecological disruptions. The realities of climate change are beginning to reveal horrific social injustice enforced by unrelenting profit-driven economies and fossil fuel industries. David Wallace-Wells (2017) believes parts of Earth will likely become close to uninhabitable by the end of this century. He argues that humanity is not alarmed enough and unable to comprehend the scope of this crisis. Climate change is no longer just a scientific issue—the world's leading scientists have clearly outlined the facts and necessary actions for decades. This is a cultural crisis that requires interdisciplinary action to mobilise communities and inspire collective action. The shift to clean energy and more sustainable lifestyles is possible, yet public engagement remains to be the greatest challenge.

Many have suggested the arts can play an active role in inspiring community action and social change. Examples such as *Liberate Tate*—which

L. Barclay (✉)
Griffith University, Brisbane, QLD, Australia
e-mail: l.barclay@griffith.edu.au

M. Droumeva, R. Jordan (eds.), *Sound, Media, Ecology*,
Palgrave Studies in Audio-Visual Culture,
https://doi.org/10.1007/978-3-030-16569-7_8

initiated a series of public live art events to successfully coerce the Tate to end its sponsorship deal with BP—clearly demonstrate the valuable role of the arts in climate action. Victoria Herrmann (2017) believes sharing stories is among the most important tools we have for climate action. She believes our current climate narratives dictate our actions, and inciting feelings of crisis and inevitable doom means we are much less likely to act. "Climate change adaptation only works when we are hopeful for the future and believe that environmentally vulnerable communities have the agency to act" (Herrmann, 2017).

Elizabeth Boulton (2016) suggests that the current scientific framing of climate change has not accounted for the importance of cultural, ontological and psychological dimensions. She draws on cognitive science, evolutionary psychology and philosophy to suggest that climate change exceeds our current cognitive and sensory abilities. Boulton calls up on artists to play a vital role in responding to climate change through sensory experiences that inspire people to engage and act. Nancy Knowlton, the Sant Chair for Marine Science at the Smithsonian's National Museum of Natural History, believes sharing stories of success, hope and optimism is one of the most effective climate action methods. While acknowledging the scale of our crisis, Knowlton (2017) believes we must celebrate success to energise the next generation and provide direction and inspiration.

Sound has incredible abilities to provide embodied sensory engagement with a place and time. Active listening can make us feel immediately present and connected to our surrounding environment. Listening to changing environments—both in situ and through virtual experiences or creative interpretations—can evoke profound interconnection and empathetic responses that have the capacity to inspire climate action. This has been the catalyst for a global movement of environmental sound artists exploring the temporal complexities of climate change, sonifying data, revealing invisible ecosystems and connecting communities to the social, cultural, ecological and political layers of changing environments through sound.

The rise of environmental sound art has also seen the emergence of artists working in a highly interdisciplinary context—initiating collaborations across sciences, activism and education, highlighting the value of sound in understanding changing environments. This resonates strongly with the field of acoustic ecology, particularly R. Murray Schafer's initial premise that we should attempt to hear the acoustic environment as music and take responsibility for its composition (Schafer, 1977).

As a composer and sound artist, I have shifted from writing music in response to the environment to designing large-scale participatory projects that encourage communities to actively listen. I have spent the last decade developing socially embedded acoustic ecology projects to inspire connection with major ecosystems across our planet through sound. These projects place equal focus on the artistic and scientific possibilities of listening and embrace emerging fields of biology measuring environmental patterns and changes through sound. I frame acoustic ecology as an accessible and dynamic interdisciplinary field concerned with the social, ecological and cultural contexts of our sonic environment. My research has drawn on emerging fields of biology including ecoacoustics, indigenous knowledge systems, sensory ecology, deep listening, psychoacoustics and responsive community engagement to position acoustic ecology as a critical interdisciplinary field to inspire climate action and understand the temporal complexities of changing environments.

Through acoustic ecology education programmes, field recording workshops, community soundwalks and a wide spectrum of artistic and scientific collaborations, I have found that sound is a powerful tool to connect and inspire communities. This expansion and diversification of my artistic practice has been rewarding, and I have seen clear value in embedding creative projects in interdisciplinary research that has social purpose and cultural and ecological relevance to a community and environment. This means my compositions and installations have become milestones in these participatory projects, as opposed to final outcomes. These large-scale projects continue to expand and evolve, and much of my recent work has focused around the interdisciplinary design of this research, highlighting the impacts that range from acoustic ecology curriculum development with remote communities in Australia to influencing policy with international programmes in UNESCO Biosphere Reserves. My documentation of these initiatives rarely dissects my personal creative outcomes, predominately because my focus shifted to working directly with local communities and my creative work felt less relevant in the bigger picture. Despite the broadened perspectives, the creative outcomes still remain at the core of these initiatives—driving the initial inspiration and acting as a catalyst through the ongoing development.

This chapter seeks to examine two creative works resulting from my research projects to highlight the influence of acoustic ecology on my personal artistic practice. These creative works are designed to demonstrate the interdisciplinary value of acoustic ecology in engaging and inspiring local and global communities to listen to the environment. The selected projects *Hydrology* and *Listening Underwater* represent different processes and approaches to working with environmental sound. While my work covers a diversity of environments, from subtropical cloud forests to temperate deserts, I have opted to focus this chapter around the acoustic ecologies of aquatic ecosystems. The two creative projects explore opportunities for acoustic ecology to continue expanding and evolving through education, engagement and interdisciplinary creativity. Prior to introducing *Hydrology* and *Listening Underwater*, I will expand on my personal perspectives on acoustic ecology and environmental field recording.

The Art and Science of Environmental Field Recording

Field recording has always been integral to my artistic practice. I work primarily with environmental sound to create immersive performances and interactive installations to bring attention and awareness to fragile ecosystems and changing environments. My fascination with field recording began in undergraduate composition studies when I first started to experiment with electroacoustic music. I used a Sony MiniDisc recorder to capture local soundscapes and was immediately struck by the possibilities of environmental sound as a tool for environmental engagement. These initial ideas were solidified during early environment projects such as *Confluence* (2005) and *Wolf Rock* (2008), which both engaged with conservation groups to use environmental field recordings and electroacoustic compositions to inspire ecological engagement.

Many of my projects have revolved around aquatic ecosystems, which led me to work with hydrophones (underwater microphones) as an essential element to my field recording process. Installations such as *Sound Mirrors* (2009–2011) drew on hydrophone recordings I collected from river system across the world. I was consistently fascinated by soundscapes beneath the surface of the water and found this unpredictable sonic world captivating. I regularly facilitated acoustic ecology workshops using hydrophones and found communities were equally enthralled by the sounds of river systems. The common narrative of rivers and oceans as quiet, tranquil environments could not be further from the truth when the hydrophones often reveal a dynamic soundscape of aquatic insects, snapping shrimp, fish and various other species.

Without a strong scientific background, I found species identification in both terrestrial and aquatic recordings challenging. While my early compositions were not intended to serve as biodiversity analysis tools, audience members would often ask about particular sound sources in the aquatic recordings and what they meant to the health of the ecosystems. This was often because the sound world of aquatic ecosystems, particularly freshwater environments, was foreign to many listeners. I was often challenged by these questions, as while I could clearly identify some sounds, such as insects or fish, there were extensive recordings with sounds I found impossible to identify. Given that the intention of my early work was to act as tools for environmental engagement and awareness, I felt it was important to develop a stronger scientific grounding for my practice. In 2011, I began a series of collaborations with biologists to learn more about the environments I was recording, particularly in the context of UNESCO Biosphere Reserves. It was this process that became the catalyst for my current interdisciplinary approaches and participatory research design that is evident in projects such as *Biosphere Soundscapes*, *River Listening* and *Sonic Reef* that house the two creative projects featured in this chapter.

In the context of aquatic ecosystems—if we look at the surface of a river, lake or ocean, it is virtually impossible to detect environmental changes. The impacts of climate change and ecological disturbances are often very visible in terrestrial environments, yet dramatic changes in aquatic ecosystems can go unnoticed simply due to visibility—we simply cannot see beneath the surface. In my early research for the *River Listening* project in 2013, I was extremely confronted by the technique that is still used today to monitor freshwater biodiversity and to measure the success of river restoration. The scientists I work with, particularly Dr. Simon Linke at the Australian Rivers Institute, believe that our current techniques such as electrofishing or capture and release do not provide accurate and reliable information. These are incredibly invasive techniques, and the resulting information does not reflect the reality of the environment. Through the *River Listening* project, we have been advocating for the use of hydrophones as non-invasive accessible tools to understand river health. Many of the species used to measure biodiversity, such as aquatic insects and fish, are soniferous, meaning it is realistic and possible to monitor the presence or absence of freshwater species with hydrophones (Linke et al., 2018).

The scientific approaches of *River Listening* are combined with creativity and digital technology to form an interdisciplinary research project exploring the cultural and biological diversity of global river systems through sound. *River Listening* examines the creative possibilities of accessible and non-invasive recording technologies to monitor river health and engage local communities in the conservation of global river systems through workshops, field recording, soundwalks, events and creative projects. *River Listening* extends and expands acoustic ecology as an inclusive, socially engaged, accessible, interdisciplinary field that can inspire communities across the world to listen to the environment and explore the sonic diversity of local rivers through various artistic and scientific approaches. I am passionate about the role of acoustic ecology in conservation and environmental management, particularly as scientists clearly need better tools for ecological monitoring. Climate change continues to cause unfathomable disruptions resulting in the rapid decline of species, which means accessible and reliable biodiversity analysis is critical to understanding changing ecosystems. While acoustic monitoring is relatively underdeveloped in freshwater environments, recent years have seen a rapid increase in non-invasive monitoring involving auditory recordings in terrestrial and marine environments. This practice has evolved from bioacoustics, a well-established field of acoustics and biology studying animal communication (Krause, 1987) and intersects with many of the foundations of acoustic ecology and more recently adopted terms soundscape ecology (Pijanowski, Farina, Gage, Dumyahn, & Krause, 2011) and ecoacoustics (Sueur & Farina, 2015).

There are striking similarities between the definition of acoustic ecology, ecoacoustics and soundscape ecology. Acoustic ecology is often misinterpreted in literature, or misrepresented as a purely humanities field, with key contributions being vocabulary for sound (Pijanowski et al., 2011). Ecoacoustics in particular has recently been coined to define a new field of science that studies sound along a broad range of spatial and temporal scales to understand environmental changes (Sueur & Farina, 2015). The expanding engagement in ecoacoustics from scientific disciplines is largely due to the accessibility and affordability of acoustic recorders and increased interest in environmental sound as a non-invasive proxy for monitoring environmental changes (Towsey, Parsons, & Sueur, 2014). Despite the clear scientific focus, ecoacoustics calls for greater collaboration with other disciplines including electronics, remote sensing, big data and social sciences (Sueur & Farina, 2015), and also advocates for creative approaches, particularly with field recordists such as Bernie Krause and David Monacchi championing the field.

What separates acoustic ecology from ecoacoustics? Many consider the terms as interchangeable, yet the recent ecoacoustics literature rarely acknowledges the history of acoustic ecology. Acoustic ecology has attracted critics, particularly when it comes to the subjective nature of some of the early terminology surrounding noise. Some scholars find the discipline restrictive; Australian sound artist Jordan Lacey has called for a reimagining of acoustic ecology and believes we should build a new relationship with urban noise, rather than attempt to escape it; we should actively reshape the 'urban roar' (Lacey, 2016).

The contentions with 'noise' are also present in the terminology associated with soundscape ecology and ecoacoustics. When the health of an environment is measured through the presences of biophony (biological sounds) and anthrophony (non-biological natural sounds), the resulting data often have negative connotations around the anthrophony (sounds produced by humans). While invasive anthropogenic noise is certainly negative in many contexts, categorising all anthrophony as negative is deeply problematic in the context of acoustic ecology projects where the social and cultural layers of the soundscapes play a vital role in the sonic environment. In my personal experience, this has been particularly problematic when working with indigenous communities whose cultural and ceremonial oral traditions are deeply connected to the environment. This has been addressed in the literature with the introduction of the term technophony, which refers to sound produced by machinery and technology. Stuart Gage advocated for this shift to acknowledge that sounds produced by humans should theoretically be considered as biophony (Dzieza, 2014).

The differences between acoustic ecology and ecoacoustics are evident in field recording approaches, particularly the role of human perception and the intended use of the resulting recordings. In the context of ecoacoustics, recordings are generally captured remotely, at multiple sites, over a broad range of spatial and temporal scales. Studies occur over long periods, to generate continuous recordings of weeks, months and, in some instances, years. These high quantities of data require acoustic indices and classifier algorithms for analysis, as it is simply impossible to listen to the resulting recordings. Many of the current algorithms and analysis software function more effectively on mono recordings, with many designed specifically for low-resolution mono recordings. The practicalities of long-durational recordings also mean monophonic and low sampling rates are more common, as the power consumption and data storage required for long-term, multi-channel, high-fidelity recordings can quickly become cost-prohibitive. As acoustic sensors become more accessible and affordable, this is beginning to change, with more scientific studies embracing higher-quality recording techniques.

Aesthetic approaches to field recording more commonly associated with acoustic ecology opt for quality over quantity. Multi-channel recording techniques such as ambisonic and surround sound microphone arrays are increasingly popular, and high sampling rates are expected. These recordings can be used for biodiversity analysis or species identification purposes, but are predominately produced for aesthetic listening experiences—both as standalone nature sound recordings and as source material for creative works. I have previously argued that this divergence relates to the role of human perception in the process (Barclay & Gifford, 2017). Ecoacoustics seeks to remove subjective perception and leverage computational advances to analyse large data sets. Acoustic ecology typically privileges human perception through activism and education in inspiring communities to listen, the foundations of the field revolve around active listening and, according to Schafer's notion, we should attempt to hear the acoustic environment as music and should take responsibility for its composition (Schafer, 1977).

Through my recent research projects *Biosphere Soundscapes* (mapping the changing soundscapes of UNESCO Biosphere Reserves) and *River Listening* (exploring the cultural and biological diversity of global river systems), we have argued that these approaches do not need to be mutually exclusive. Acoustic indices and classifier algorithms are still in a phase of development, and the most effective machine listening advances cannot compare to the abilities of human auditory perception in situations such as pattern recognition. Algorithms exclude critical elements of the soundscape, and existing ecoacoustic analysis techniques operating in time or frequency domain may have major limitations that are insensitive to the dynamic patterns of interaction (Eldridge et al., 2016). These large-scale acoustic ecology projects advocate for interdisciplinarity in field recording that embraces both perspectives with resulting recordings that have value across arts and sciences.

The *Biosphere Soundscapes* residency and workshop programmes I have developed have called for greater cross-fertilisation and interdisciplinary dialogue around existing field recording practices. These programmes have advocated for acoustic ecology to be the most appropriate lens for exchange, exploration and expansion. We invite artists and scientists to collaborate in designing long-durational remote recording strategies and participate in onsite recording sessions using high-quality equipment where perspectives on listening, perception, microphone placement and intention are shared. The recordings are used for both artistic and scientific purposes, and the process itself consistently results in valuable dialogue where aesthetic and scientific approaches to listening and analysis influence the development of ongoing projects. This ranges from artists inspiring scientists to use different techniques in selecting recording positions and locations to critical enquiry into the future possibilities of acoustic ecology.

In parallel to the in situ field recording sessions during *Biosphere Soundscapes* residences and workshops, we facilitate remote recording studies, usually with microphones arrays that are installed for long periods of time, ranging from one week to three months in the workshop format. These studies are facilitated by scientists who specialise in ecoacoustics and often reveal the methodological process does not draw on the value of our auditory perception. While the scientific advancements of ecoacoustics have provided valuable ecological information and clear evidence on the validity of the research, listening has not played an active role in many existing studies. Acoustic sensors are often deployed in the field by scientists at predetermined locations and analysed using automated algorithms. The resulting data are produced without drawing on the far more nuanced capacity of our auditory perception. Yet listening and our comprehension of the acoustic data can inform every stage of the process and dramatically increase our abilities to collect precise information—from selecting locations in the field to microphone techniques and analysis. Musicians and composers who have had extensive aural training have exceptional abilities to recognise patterns and changes through sound. The specialist auditory training that composers possess allows them to process and organise vast amounts of acoustic data that can play a valuable role in advancing the scientific possibilities of ecoacoustics or acoustic ecology. This collaboration between artists and scientists during *Biosphere Soundscapes* projects is almost always mutually beneficial; scientists developing more appreciation and understanding in our listening can enhance their research methods, and artists can develop a deeper understanding of ways to use sound to measure ecological health.

While potentially contentious, I prefer to consider acoustic ecology as the umbrella field that is constantly evolving in exploring the social, cultural and ecological aspects of our environment through sound—a truly transdisciplinary field embracing the artistic and scientific possibilities of listening to the environment. In my keynote presentation at the *2017 Sound + Environment Conference* at the University of Hull, I used the following slide to demonstrate these possibilities (Fig. 8.1).

This slide represents my personal perspectives on the possible intersections and transdisciplinarity of acoustic ecology and is not intended to be interpreted in a hierarchical way. I am not suggesting that acoustic ecology is more important than ecoacoustics, just that it has the scope and capacity to connect and draw on elements from many of these divergent fields. From cultural to biological, intimate to remote, acoustic ecology considers the aesthetic properties of sound and personal perception, but also the notion of sound as data and a measure for environmental health. I strongly believe the Deep Listening practices of Pauline Oliveros and the notions of Steven Feld's acoustemology with sound as a distinctive medium for knowing the world (Feld, 1996) are integral to acoustic ecology. Yet I am equally passionate about the scientific advancements in ecoacoustics that advocate for sound to play a critical role in policy development and environmental management (Deichmann et al., 2018). Accessible and affordable acoustic sensors are "revolutionising environmental monitoring and leading to a new thrust in environmental research and education" (Farina

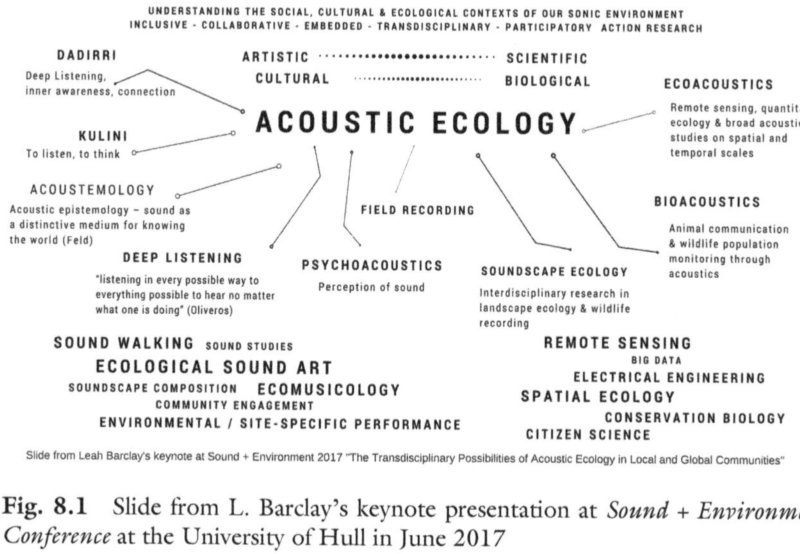

Fig. 8.1 Slide from L. Barclay's keynote presentation at *Sound + Environment Conference* at the University of Hull in June 2017

& Gage, 2017). It is perhaps one of the most exciting times to be working at the intersection of sound and ecology, with developments across biological sciences, computational technology, engineering, big data and humanities all exploring the way we can use sound to understanding changing environments. This provides an exceptional opportunity for acoustic ecology to embrace these interdisciplinary possibilities.

I believe this interdisciplinary and somewhat evolving framework is exciting and potentially necessary in the context of acoustic ecology community projects designed to truly explore the social, cultural and ecological contexts of our environment through sound. My creative work takes a similar trajectory—my personal approach to field recording is for both artistic and scientific purposes, but I see all of the recordings as possible source material in the creative works. The resulting performances and installations are designed to encourage listeners to engage with ecosystems and facilitate embodied connections with the environment. I work with the materials intuitively and draw on climate research to construct sonic narratives that focus on hope for our collective future while highlighting the fragility of rapidly changing ecosystems. These works are always embedded in research projects that provide direct pathways for listeners to take actions and connect with conservation organisations associated with the project. In some instances, my compositions feature pure sound recordings, while others undergo extensive processing to draw out particular qualities of the soundscape—each project evolves in response to the environment. The remainder of this chapter introduces two examples of creative projects resulting from my acoustic ecology research in aquatic environments. These examples outline creative projects that demonstrate my trajectory of engagement with acoustic ecology through immersive compositions that encourage deep listening to ecosystems with strong foundations in scientific research.

Hydrology: Sound Installation

Hydrology is an immersive sound environment exploring the value of listening to aquatic ecosystems that cover over 70% of Earth's surface. The work draws on a decade of my hydrophone recordings from projects including *Sound Mirrors* (2009–2011), *The DAM(N) Project* (2010–2013) and my current major research projects *Biosphere Soundscapes*, an interdisciplinary venture exploring the changing soundscapes of UNESCO Biosphere Reserves, and *River Listening*, a project which examines the artistic and scientific value of freshwater soundscapes. *Hydrology* was commissioned by Professor Sabine Breitsameter as part of her 3D audio research project at Hochschule Darmstadt University of Applied Sciences in Germany. This piece was commissioned to accompany my keynote presentation at the International Symposium *Disseminations - Media Cultural Practices for the Digital Age* at Museum Schloss Fechenbach in Dieburg, Germany, in December 2016.

Professor Breitsameter's 3D audio research is focused around object-based composition using SpatialSound Wave (SSW) technologies developed by the Fraunhofer Institute for Digital Media Technology (IDMT) in Germany. The Fraunhofer Institute is renowned for developing groundbreaking technologies in audiovisual media and leading applied research in spatial sound. SpatialSound Wave (SSW) is an innovative software system that allows composers to work with audio files as sound objects and distribute the material with a specialised process of wave field synthesis. This technology draws on principles of wave propagation to create highly immersive and realistic three-dimensional sound environments. This object-based system allows you to interact with a virtual sound field and reproduce spatial sound with remarkable clarity in various speaker configurations. I was particular attracted to the idea of flexible speaker arrays above the audience as I find standard spatial audio (5.1 and 7.1) difficult to produce truly immersive listening experiences, particularly with aquatic recordings.

Hydrology was composed for the Hochschule Darmstadt University of Applied Sciences SSW system during my residency over the course of one week. While this was an extremely tight production schedule, I had spent a considerable amount of time developing the source materials and preparing the structure of the piece prior to arriving in Germany. I had selected a collection of field recordings from freshwater and marine ecosystems across the planet and worked extensively on composition sketches. The selected recordings were collected during various interconnected research projects, and each represented a moment in the field where I was enthralled with the acoustic ecologies beneath the surface of the water. I wanted the piece to demonstrate the incredible sonic diversity of aquatic environments and offer audiences the opportunity to experience these soundscapes as if they were listening underwater.

The locations covered 16 ecosystems including coastal mangroves in the Sian Ka'an Biosphere Reserve in Mexico, Queensland's iconic Great Barrier Reef and frozen rivers in Norway. The recordings had all been created in explorations of the scientific possibilities of aquatic ecoacoustics and included snapping shrimp and aquatic insects in Australian rivers, rare dolphins in the Central Amazon and humpback whales migrating through the Pacific Ocean. These species can all be indicators of ecosystem health, and while I did not want the piece to feel like a bioacoustics database, I wanted to convey the value of listening for both artistic and scientific purposes. *Hydrology* was responding to the idea that the hydrophones provide access to another world of exploration and a non-invasive way of understanding changing aquatic ecosystems. The work reveals the soundscapes beneath the surface of aquatic ecosystems we might not traditionally think about (Fig. 8.2).

Each recording I selected evoked a particular memory of the environment, and I processed the material to create a series of short soundscapes that would form the foundation for the composition. This resulted in

Fig. 8.2 Artist Leah Barclay recording underwater with hydrophones

approximately one hour of audio that was prepared prior to beginning the composition process in Germany. SpatialSound Wave allows composers to create three-dimensional sound objects and position these in any spatial arrangement. The software provides an accessible visual interface where it is relatively easy to move sound throughout the environment. Each sound object can have independent input signals, spatialisation and processing, and the software environment truly feels like sculpting sound. The authentic feeling of immersion is due to the SpatialSound Wave system computing each sound object signal in real time through the defined speaker array with intelligent signal processing. My first experiments with the software were invigorating, and while I had a clear idea of the composition prior to arriving, the flexibility and intuitive design of the system certainly influenced my creative approach. The visual interface allowed me to immerse the venue in aquatic recordings and sculpt audio objects floating through space as if we were listening on the floor of the ocean. There were 32 speakers used for diffusion, and while multi-channel composition has always been integral to my practice, this was one of the first experiences where I was able to intuitively design realistic acoustic ecologies that were not restricted with speaker placement.

I initially experimented with sending sound objects to the entire array of 32 speakers to create a realistic aquatic ecosystem. I sculpted and composed the soundscapes in the three-dimensional environments and transitioned through the diversity of acoustic ecologies featured in my field recordings. As I was exploring the scientific possibilities of aquatic ecoacoustics, many of the recordings of marine and freshwater species were left untreated and positioned throughout the soundscapes with realistic spatial gestures. As an example, I was able recreate a pod of Amazon River dolphins circling the listeners at various distances. My research has consistently focused around the value of realistic sonic embodiment in ecological engagement, and it was clear this method of working with object-based spatial composition offered remarkable possibilities to create compelling sonic environments.

Hydrology was installed as part of the ongoing 3D sound exhibition during the *Disseminations* symposium and received positive feedback from the audience. I was particularly struck by the fact that many local students were unaware that freshwater and marine species were capable of making such dynamic sounds. I always make a point that my compositions are accessible and offer them back to the collaborating communities and conservation organisations for ongoing engagement and awareness projects or campaigns. The only disadvantage of working with the SSW software and diffusion system was the ongoing accessibility. There are very few performance venues in the world equipped with the appropriate technical capabilities, and purchasing or hiring the equipment is not financially realistic for most projects. As with many of my creative works, *Hydrology* became a multi-platform composition, with various outcomes to ensure the accessibility and ongoing engagement.

I created a basic stereo version to share with the collaborating communities and conservation organisations; these included community groups associated with UNESCO Biosphere Reserves and river restoration organisations in Australia. I also developed a multi-channel version that has featured at various events including the 2017 World Science Festival in Brisbane, Australia. This version was composed for eight-channel surround sound and constructed using very similar spatialisation to the original version in Germany (without the additional speakers for height). While the ring of eight speakers surrounding the audience was certainly immersive, I believe the feelings of true immersion were more prevalent using the SSW system where sound felt like it was freely flowing above and around the listeners. In the initial version, audience members commented that the sound felt like it had the properties of water and this was difficult to recreate without the SSW technology.

In addition to the *Hydrology* spatial compositions, I created an augmented reality soundwalk that mapped the soundscapes throughout an urban environment. This was building on my existing research into the value of mobile technologies and locative audio as a tool for environmental engagement (Barclay, 2017). The *Hydrology* soundwalk was launched as part of the 2017 World Science Festival, and the soundscapes were positioned along the Brisbane River and throughout the Southbank Parklands in Queensland, Australia. Listeners downloaded a free app and triggered audio files with GPS as they walked throughout the parklands. This adaptation of the work required some compromises with sound quality as all of the files were produced in a stereo format for headphone playback. The positioning of the sounds throughout the environment attempted to mirror the spatialisation, yet the experience would be different for every listener based on their position and movement. I designed the soundwalk as a non-linear installation, but still invested considerable time in recreating the fluidity of the work regardless of the listener's direction and pace. The *Hydrology* walk featured 40 individual soundscapes that each had accompanying images and text. This provided options for the audience to simply walk and listen, or explore the aquatic recording at a deeper level. This was particularly valuable in the context of a science festival where listeners were immediately curious about species identification and recording locations.

The *Hydrology* soundwalks shared exactly the same intention and source material as the original composition but were a completely different realisation and experience. The immersive sound environment created with the SpatialSound Wave system allowed me to sculpt sound objects in space and immerse listeners in the sonic complexities of aquatic ecosystems. I am passionate about the value of spatial audio and believe these technologies are the most effective in inspiring embodied ecological connection in listeners. However, my commitment to accessibility always leads me to create multi-platform outcomes, and the *Hydrology* soundwalks attracted listeners who would unlikely attend an electroacoustic concert. The soundwalks were accessible to young children, and the flexibility of the listening experience meant they could listen to just one sound or explore different parts of the walk throughout the day. Children as young as three were enthralled by the sounds of snapping shrimp and cat fish in the Brisbane River, and it was clear that the mobile listening experiences extended the engagement of the project. As an example, the *Hydrology* concert at World Science Festival was attended by 75 people (the venue capacity), and the soundwalks were experienced by over 450 listeners throughout the event. Providing different pathways for audiences to access the works has been integral to my research in exploring the creative possibilities and accessibility of acoustic ecology.

Hydrology draws on emerging fields of science and explores the diverse sonic properties of water to reveal the acoustic ecologies beneath the surface of oceans, lakes and rivers across the planet. The piece is realised in various formats and approaches sound as a fluid material that is sculpted in space. This work has connected outcomes and recording databases from multiple projects and has been effective in engaging listeners in aquatic environments. *Hydrology* will continue expanding as both a live performance and installation, and the next steps will involve collaborations to measure the impact. As the intention of *Hydrology* is to influence ecological connection and engagement, the next phase of this research will involve developing and applying a critical evaluative framework to measure the ecological engagement and behavioural change facilitated through embodied sonic experiences. I believe these experiences can lead to attitudinal and behavioural changes, but there has been limited investigation on the role of the auditory perception in these experiences.

There is an incredible body of environmental sound art emerging in response to climate change (Gilmurray, 2017). However, no consensus has emerged regarding effective technical and aesthetic approaches, and no evaluation framework exists for assessing the ecological engagement or intended behavioural change afforded by various approaches. Two new iterations of *Hydrology* (as an immersive performance and as a virtual reality installation) are being used as a case study to work in collaboration with a team of cognitive scientists and behavioural psychologists to investigate the impact, affect and cognitive change in response to embodied sonic experiences. This study is providing greater insight into how these experiences are processed by individuals and how ecological sound art and acoustic ecology can motivate and inspire listeners to engage in climate change mitigation and adaptation. This interdisciplinary research in embodied cognition represents another example of the transdisciplinary possibilities of acoustic ecology.

Listening Underwater: Live Performance

Listening Underwater was a live performance for the 2017 Smithsonian Earth Optimism Summit in Washington, DC. The event celebrated a change in focus from problem to solution in the area of global conservation, with an unprecedented gathering of thought leaders, scientists, environmentalists, artists, civic leaders and international media. I was invited to give a presentation on acoustic ecology and to perform live to open the session themed around science, conservation and inspiration. I wanted the performance to convey a sense of fascination with listening underwater and also highlight the artistic and scientific possibilities of environmental sound. I decided to incorporate real-time streaming hydrophones with a live mix of field recordings from various aquatic environments. The hydrophones were streamed live from a humpback whale breeding location in Puako, Hawaii, courtesy of the Jupiter Research Foundation.

This performance was inspired by my long-term fascination with live streaming audio and the exciting artistic and scientific possibilities that are emerging with this technology. In 2009, during my preliminary doctoral research, I came across a link for PALAOA, the Perennial Acoustic Observatory in the Antarctic Ocean (PALAOA is the Hawaiian word for whale). This website featured a live audio link of hydrophones in Antarctic on the Ekström Ice Shelf, and I was immediately captivated by this incredible soundscape. I listened continuously for several hours and could hear

Weddell seals, killer whales and a range of cetaceans passing the hydrophones. The fact that the stream was live added fascinating element to the listening experience that was immediately compelling. The PALAOA stream was established for scientific research, but had an array of artistic possibilities. It was launched in 2005 on the basis of using passive acoustics to detect marine mammals. The onsite recording and live stream ran almost continuously until 2014 to study the acoustic repertoire of whales and seals in an environment that was virtually free from anthropogenic noise (Kindermann et al., 2007). The stream drew from four hydrophones located through boreholes below the 100-metre-thick Ekström Ice Shelf. From a scientific perspective, the resulting data were successfully used to detect species, map migration patterns and investigate the potential impact of anthropogenic sound such as distant shipping traffic (Kindermann et al., 2007).

I did not personally use the PALAOA stream in artistic projects, though it was my initial inspiration for various live streaming initiatives that have occurred through both *River Listening* and *Biosphere Soundscapes*. The pilot for *River Listening* was launched in 2014 by exploring the creative and scientific possibilities of live streaming hydrophones in global river systems. On World Water Day 2014, we established a live stream in the Brisbane River in Australia and connected with a node in New York City along the Hudson River to initiate a discussion about the value of aquatic acoustic ecology as part of the World Water Day Symposium in New York. The initial phase of *River Listening* in Australia involved community streaming workshops across Queensland, and the project officially launched on the iconic River Thames in London during the 25th anniversary of the EVA London International Conference in June 2014. This involved hosting a seven-day live stream with hydrophones installed on the HMS Belfast, permanently moored on the River Thames. The intensity of the Thames stream, reminiscent of a busy highway, was compared to our database of river recordings in Australia and demonstrated the differences in aquatic sound propagation and the potential value of live streaming audio. We used social media, particularly Twitter, to facilitate an online conversation around the stream, and the project became a valuable learning experience to design the ongoing project.

Many of the performances and installations resulting from *Biosphere Soundscapes* and *River Listening* have incorporated live streams that we have established, but in all cases these have been temporary streams running for periods varying from several days to three weeks. The infrastructure required for permanent streams can be cost prohibitive in many research projects, particularly streaming from remote locations in Australia where the data requirements alone can be extremely expensive. As an example, a basic Mp3 encoded stereo hydrophone stream running at 320 kbps will use 144 megabytes per hour, 3.5 gigabytes per day, over 100 gigabytes per month and at least 1.3 terabytes per year. The most affordable 4G mobile broadband in Australia for regional areas can be calculated at $150 per 25 gigabytes, meaning the data costs for the stream are approximately $600 per month. The various technical challenges and global opportunities for live streaming audio have resulted in a culture of collaboration emerging through artistic and scientific networks working in this field. This was the foundation for my collaboration SoundCamp, a London-based organisation coordinated by Grant Smith, Maria Papadomanolaki, Kirsty Collander Brown and Dawn Scarfe.

I discovered the work of SoundCamp through *Reveil*, a 24-hour live audio stream and radio broadcast that tracks sunrise around the globe, relaying live audio from open microphones across the world. SoundCamp is a platform for artistic and transdisciplinary work in the emerging field of live streaming audio and work in partnership with organisations and collaborators to research open-source solutions for live streaming and establish permanent streams as a resource for artistic and scientific purposes. The *Reveil* project is a wonderful example of combining the possibilities of acoustic ecology and emerging technologies—the notion of listening continuously to dawn unfolding across the globe is poetic and accessible, yet also highly innovative and reminiscent of the *Environmental Radio* proposition by Bruce Davis to install microphones in wilderness areas and transmit natural soundscape to urban listeners (Schafer, 1977). The *Reveil* project shares a sense that remote listening can shift our sense of perspective, but it does not prefer 'hi-fi' rural to 'lo-fi' urban sounds as in traditional acoustic ecology (SoundCamp, 2018).

The *Reveil* streams typically come from inhabited places and feature a diversity of environments. The project is developed in collaboration with Locus Sonus and widely accessible online through the Locus Sonus Sound Map, which has been active since 1995 as a collaborative online map hosting live audio streams (SoundCamp, 2018). In addition to facilitating community streaming projects, SoundCamp and Locus Sonus have partnered with various organisations to connect existing streams. This includes the Pauko stream I featured in the Earth Optimism *Listening Underwater* performance. While I had heard the Puako Humpback whale streams prior to working with SoundCamp, it was our partnership that facilitated my initial collaborations with the Jupiter Research Foundation to begin working with their hydrophone streams in a creative context.

The Jupiter Research Foundation was established in 2003 as a non-profit scientific research organisation dedicated to research and development around new technologies for monitoring natural environments. The Humpback Whale Project was their first major research initiative, with an objective to stream live hydrophones for research and public engagement. They have streamed the humpback whale migration season annually since 2003 and been engaged in a constant process of research and experimentation to improve and optimise their technologies. The initial stream featured solar-powered radio buoys with wireless transmitters with the hydrophones submerged at a depth of 18 metres. The research team now use more robust Wave Gliders that allow the hydrophones to shift positions and transmit live audio using cell phone signals (Jupiter Research Foundation, 2017). The stream is active for several months per year, usually from December to April when the whales are in Hawaii mating and birthing calves. The songs of male humpback whales have attracted extensive research and have been a source of inspiration for a large body of musical works. Conservationist Roger Payne is widely credited for contributing towards the global conservation of humpback whales after his recordings were included in a 1979 edition of National Geographic distributed to millions of people. While there had been previous scientific research, Payne brought humpback whale song to the general public through his 1970 release *Songs of the Humpback Whale* which became the highest selling environmental recording and a milestone in the marine conservation movement (O'Dell, 2010).

The fascinating structures and complexity of humpback whale song have inspired extensive research related to environmental changes, much of which I have been actively engaged with through the *Biosphere Soundscapes* project. This has been particularly prevalent with our work in the Great Sandy Biosphere Reserve in Queensland which is a migration zone for Eastern Australian humpback whales. For the performance of *Listening Underwater*, I sampled recordings of humpback whales from the east coast of Australia and mixed them live with freshwater and marine recording from the coast of Mexico and Hawaii. The performance was designed to be completely live in response to the stream, with the real-time audio from Puako guiding the composition and performance. Naturally there are extensive risks involved in such performances, and having worked with the Puako stream in a live context on several prior occasions, I was aware of the unpredictable nature that has been both incredible and potentially disastrous. In an earlier concert in 2015, I was improvising with the Puako stream when a sudden interference distorted the hydrophones and halted the performance. The success of the work was also reliant on stable internet connections in the venue to host the stream so there were a number of technical components to consider when preparing this work.

The performance of *Listening Underwater* in Washington was an exciting opportunity at a high-profile Smithsonian event, so I developed various scenarios to mitigate potential technical issues. These included a second computer running the stream with a backup internet connection and some basic tools within my software environment that allowed me to fade or mute the stream if distortion or interference became an issue. Fortunately, the clarity of the hydrophones was perfect, and multiple whales were singing in close proximity to the hydrophones. I started the stream as people entered the venue and diffused it subtly throughout the speaker array along the ceiling; this performance gradually sculpted sonic textures throughout the venue that crossed various marine environments responding to the real-time audio stream from Puako. The audience responded positively, and my presentation that immediately followed the performance allowed me to introduced the technology and unpack the process, including a detailed introduction of my approaches to acoustic ecology and the scientific foundations that underpin my creative work.

The Earth Optimism Summit attracted an interdisciplinary audience who were predominately unfamiliar with acoustic ecology. This was a fantastic opportunity to provide an accessible introduction and outline some of the future possibilities of the field. I find non-specialist events fantastic platforms to expand the interest and engagement in acoustic ecology, and given the intention of my research, I feel a responsibility to present my work outside of traditional academic circles. *Listening Underwater* is an integral part of my ongoing acoustic ecology research, and the foundations of this performance have contributed towards the creative development for the latest *Biosphere Soundscapes* initiative—*Biosphere Open Microphones (BIOM)*.

BIOM is a multi-partner acoustic ecology project establishing a network of live streaming sites in UNESCO Biosphere Reserves for public engagement and interdisciplinary research. The project brings together a group of pioneering arts and science initiatives to establish new live streams from places of exceptional ecological and acoustic interest that will feature on an accessible live soundmap. BIOM builds on the existing work of *Biosphere Soundscapes* in collaboration with Locus Sonus (Marseille, France), Cyberforest (Tokyo, Japan), Jasper Ridge Biological Preserve (Stanford, USA), SoundCamp (London, UK), Streaming Museum (New York) and the World Network of UNESCO Biosphere Reserves (Paris).

The project formally launched during the *2017 Sound + Environment Conference* at the University of Hull and featured a series of pilot streams including existing sites in Japan and new sites in the North Devon Biosphere Reserve (Plymouth, UK) and Noosa Biosphere Reserve in Australia. BIOM will provide a network for research, monitoring, artistic work and community engagement through a real-time acoustic observatory reflecting the diversity of terrestrial and aquatic biomes in UNESCO biosphere reserves across the world. The project will focus on community engagement and partner with local organisations to develop and maintain the streams at each location. The audio will be recorded and maintained at the University of Tokyo to provide long-term datasets, and the BIOM team will develop and facilitate acoustic ecology programmes specific to each location. Our fieldwork will revolve around public engagement; outdoor gatherings are opportunities for new publics and volunteers to engage with acoustic ecology—locally and remotely. As with *Biosphere Soundscapes*, the community involvement in the setup and maintenance of the streams will make them a hub for learning, collaboration, innovation, citizen science and interdisciplinary creativity.

The next phase of development will include a hydrophone stream in the Great Sandy Biosphere Reserve near Fraser Island that will run in alternate months to the Puako stream in Hawaii. The development of this stream is directly inspired by the *Listening Underwater* performance and will contribute towards a continuous live stream for humpback whale migration patterns throughout the world's oceans on the BIOM soundmap. While in very early stage, BIOM has the capacity to be a valuable acoustic observatory for a wide spectrum of artistic and scientific purposes in addition to acting as an accessible tool for engagement, activism and education around acoustic ecology.

Conclusion

The projects featured in this chapter are two examples from a large body of interconnected creative work that is embedded in participatory acoustic ecology projects that value artistic, aesthetic and scientific perspectives. This research highlights the interdisciplinary possibilities of ecological sound art and frames acoustic ecology as a socially engaged, inclusive, accessible, interdisciplinary field that can inspire communities across the world to listen to the environment. I truly believe that participatory interdisciplinary projects working directly with local communities can have significant impact, and after over of decade of embedding acoustic ecology projects in communities, it is rewarding to witness tangible changes. Despite the catastrophic impacts of climate change and the unnerving predications for the future, I am constantly inspired by the incredible work happening in acoustic ecology from both scientific and artistic perspectives and believe interdisciplinary collaborations can play a critical role in climate change mitigation and adaptation.

These interconnected projects are designed with a sense of optimism about the future and focus on facilitating connections between art, science and technology to mobilise local and global communities to listen, connect and collaborate. The emerging scientific interest in acoustic ecology feels like a new wave of engagement for the field as new discoveries are being made every day about the sonic complexity of the nature world. The rapid advancements of acoustic technologies means it is becoming more accessible to map, document and understand changing environments through sound. I see environmental sound art as an ongoing catalyst in this process, and I am optimistic that the interdisciplinary possibilities of acoustic ecology can contribute towards shifting the climate narrative to one of hope, engagement and ultimately action.

References

Barclay, L. (2017). Augmenting Urban Space with Environmental Soundscapes and Mobile Technologies. *Soundscape – The Journal of Acoustic Ecology, 16*(2), 21–34.

Barclay, L., & Gifford, T. (2017). The Art and Science of Recording the Environment. *Leonardo*. https://doi.org/10.1162/LEON_a_01517

Boulton, E. (2016). Climate Change as a "Hyperobject": A Critical Review of Timothy Morton's Reframing Narrative. *WIREs Climate Change*, 7, 772–785. https://doi.org/10.1002/wcc.410

Deichmann, J., Acevedo-Charry, O., Barclay, L., Burivalova, Z., Campos-Cerqueira, M., d'Horta, F., et al. (2018). It's Time to Listen: There Is Much to Be Learned from the Sounds of Tropical Ecosystems. *Biotropica*. https://doi.org/10.1111/btp.12593

Dzieza, J. (2014). *Scientists Are Recording the Sound of the Whole Planet*. Retrieved from https://www.theverge.com/2014/8/28/6071399/scientists-are-recording-the-sound-of-the-whole-planet

Eldridge, A., Casey, M., Moscoso, P., & Peck, M. (2016). A New Method for Ecoacoustics? Toward the Extraction and Evaluation of Ecologically-Meaningful Soundscape Components Using Sparse Coding Methods. *PeerJ, 4*, e2108. https://doi.org/10.7717/peerj.2108.

Farina, A., & Gage, S. (2017). *Ecoacoustics: The Ecological Role of Sound*. Oxford, UK: Wiley.

Feld, S. (1996). Waterfalls of Song: An Acoustemology of Place Resounding in Bosavi, Papua New Guinea. In S. Feld & K. H. Basso (Eds.), *Senses of Place*. Santa Fe, NM: School of American Research Press.

Gilmurray, J. (2017). Ecological Sound Art: Steps Towards a New Field. *Organised Sound, 22*(1), 32–41. https://doi.org/10.1017/S1355771816000315

Herrmann, V. (2017, July 12). Doomsday Narratives About Climate Change Don't Work. But Here's What Does. *The Guardian*. Retrieved from https://www.theguardian.com/commentisfree/2017/jul/12/doomsday-narratives-climate-change-dangerous-wrong

Jupiter Research Foundation. (2017). Retrieved from http://jupiterfoundation.org/

Kindermann, L, Boebel, O., Bornemann, H., Burkhardt, E., Klinck, H., Van Opzeeland, I., et al. (2007). *A Perennial Acoustic Observatory in the Antarctic Ocean*. Computational Bioacoustics for Assessing Biodiversity: Proceedings of the International Expert Meeting on IT-Based Detection of Bioacoustical Patterns, December 7th until December 10th, at the International Academy for Nature Conservation (INA), Isle of Vilm. hdl: 10013/epic.30936.d001

Knowlton, N. (2017). Doom and Gloom Won't Save the World. *Nature, 544*, 271. https://doi.org/10.1038/544271a

Krause, B. (1987). Bioacoustics, Habitat Ambience in Ecological Balance. *Whole Earth Review, 57*, 14–18.

Lacey, J. (2016). *Sonic Rupture: A Practice-led Approach to Urban Soundscape Design*. New York: Bloomsbury.

Linke, S., Gifford, T., Desjonquères, C., Tonolla, D., Aubin, T., Barclay, L., et al. (2018). Freshwater Ecoacoustics as a Tool for Continuous Ecosystem Monitoring. *Frontiers in Ecology and the Environment*. https://doi.org/10.1002/fee.1779

O'Dell, C. (2010). *Songs of the Humpback Whale (1970) Library of Congress*. Retrieved from https://www.loc.gov/programs/static/national-recording-preservation-board/documents/humpback%20whales.pdf

Pijanowski, B. C., Farina, A., Gage, S. H., Dumyahn, S. L., & Krause, B. (2011). What Is Soundscape Ecology? An Introduction and Overview of an Emerging New Science. *Landscape Ecology, 26*, 1213–1232. https://doi.org/10.1007/s10980-011-9600-8

Schafer, R. M. (1977). *The Tuning of the World*. Montgomery, AL: Random House Inc. Reprinted as *The Soundscape: Our Sonic Environment and the Tuning of the World*. Rochester, VT: Destiny Books, 1994.

SoundCamp. (2018). The Live Audio Archive. *Leonardo*. https://doi.org/10.1162/leon_a_01536

Sueur, J., & Farina, A. (2015). Ecoacoustics: The Ecological Investigation and Interpretation of Environmental Sound. *Biosemiotics*. https://doi.org/10.1007/s12304-015-9248-x.

Towsey, M., Parsons, S., & Sueur, J. (2014). Ecology and Acoustics at a Large Scale. *Ecological Informatics, 21*, 1–3.

Wallace-Wells, D. (2017). *The Uninhabitable Earth*. Retrieved from nymag.com/intelligencer/2017/07/climate-change-earth-too-hot-for-humans.html

CHAPTER 9

Listening to Renewable Energy Technologies

Linda O Keeffe

Introduction

The chapter explores field trips taken between 2015 and 2017, recording vastly different soundscapes; the first, the southern region of Iceland, the second, the Northern Terra Alta region of Spain. The trip to Iceland took place in June 2015 and was followed by two trips to Spain in July 2015 and August 2016. It was during the field trip to Iceland that I began to explore the sonic impact of renewable energy technologies within various acoustic territories. When exploring the soundscapes of renewable energy technologies, I was surprised to discover the perceptible sound levels were quite high. In addition, there was a noticeable socio-economic and cultural impact as a result of the emergence of renewable energies in certain regions. This particular impact was documented through a listening practice, which found a marked reduction of population sounds and/or a reduction in traditional economic practices, which could be defined as archetypal or historic sounds.

It is assumed that the design of these technologies is influenced and shaped by ethical considerations for the landscape. Further, the process of

L. O Keeffe (✉)
Edinburgh College of Art, The University of Edinburgh, Edinburgh, UK
e-mail: lokeeff@exseed.ed.ac.uk

M. Droumeva, R. Jordan (eds.), *Sound, Media, Ecology*,
Palgrave Studies in Audio-Visual Culture,
https://doi.org/10.1007/978-3-030-16569-7_9

developing these technologies is grounded in holistic concerns, a gestalt view where the environment and surrounding ecosystems, including human ecosystems, are considered within impact studies. This chapter explores how the development of a listening methodology allowed for a deeper understanding of the potential impact of a technological system within these two distinctive soundscapes.

Energy and the Landscape

Sustainable energy is increasingly defined as the only way forward for a planet suffering from environmental harm due to polluting energy sources such as coal, gas and oil. Pollution is often defined in terms of physical materials, which infect all ecosystems with problematic gases or chemicals, through processes of manufacturing and/or extraction.[1] Considering the environmental impact of these energy sources, it can seem problematic to question potential side effects caused by renewable energy technologies or infrastructures. However, it is our history of not examining the long-term impact of new technological systems which has led to our current environmental problems. Through auditory observation, researchers can become aware of the very different and subtle ways in which structures that harness and produce energy might begin to shape or alter social and environmental ecologies. Rachel Carson (2000) posited that man's poor management of land would lead to what she called the Silent Spring, a process whereby the soundscape of nature starts to disappear as habitats become uninhabitable. The loss of the natural soundscape would be the first fundamental shock to the system and would herald the decline of the natural environment (Whitehouse, 2015). However, because listening is not foregrounded in environmental science, we would fail to take notice of these silences until the physical and visible side effects appeared.

The emergence of the field of acoustic ecology has led to a significant number of findings framed within an audible methodology, that is, recording and analysing the sounds produced in particular environments and the impact these sounds have on a natural ecosystem. Listening to nature is now encouraged and even advanced by acoustic ecologists such as Leah Barclay (2017) and organisations such as the World Forum for Acoustic Ecology and the Acoustic Ecology Institute. A growing movement of environmentally aware sound artists and field recordists has documented the impact of noise pollution and polluting technologies within various biospheres. Artist researcher Anja Kanngieser posits that sound allows us to hear the political and socio-economic links "between humans, environments and systems of governance".[2] However, by incorporating transdisciplinary processes for soundscape examination, we have the ability to transcribe in detail what is happening within complex biological and social systems.

Listening as Practice: A Methodology

During the three-year period of research from 2014 to 2017, I developed two modes of listening, and both involved uniquely different approaches to documenting and understanding a soundscape. These have been shaped by previously developed methods for soundscape research concerning the impact of shifting urban soundscapes on communities (O Keeffe, 2015a, b). Epstein (2009) states that the key component of the field of acoustic ecology is its hybridisation of interdisciplinary problem solving. In listening to our environment, we must not only explore such concerns as noise and its impact on certain environments but also include an examination of the social experience of sound, the cognitive and psychological impact of different types of listening experiences and the aesthetics of environmental sounds.

First, the process of documenting the landscape becomes a sort of soundmapping method, whereby each heard space would be noted by time, weather conditions, a linking of sound effect to production, the position of the listener relative to the sound and a catalogue of what is seen and sensed other than sound. The use of the word mapping implies a flattening of the perceived space, a traditional cartographic approach, which de Certeau (1988) argues ignores the subjective experience of place. Instead, I created an assemblage of data from note taking to audio, video and photographic documentation, non-prescriptive open-ended interviews and participant observation. This approach involves a form of embodied listening where one tries to interpret what role sound plays in the environment and within communities, including how some sounds impact on both social and natural spaces. When thinking of a space in three-dimensional terms, it is important to examine the entire ecosystem from the sounds produced by economic practices, contemporary, traditional or lost; community and cultural sounds, natural environmental sounds, both geophonic and biophonic, and the stories told by locals. It is also important to reflect on the listeners' experience of a space, what connections and knowledge they may have about the space, to assess one's subjectivities. This mode of listening is shaped by an awareness of the social shaping of land and community structures as well as the method of attentive and observational listening.

Second, using the Oliveros (2005) Deep Listening method, it is possible to develop a listening practice whereby passive and active listening, or directional and focused listening, allows an attentive listener tune sound in or out, or manage and comprehend past sounds, making links between what is heard and what is remembered. Passive listening in this instance becomes a process of being immersed in a current sound environment while reflecting on previous sensory moments whether this was a story collected or previous listening moments. Then note taking and/or dictating the experience becomes a key tool to support the documentation of sonic moments. In this way, the act of listening becomes as important as the process of recording sound for documentary purposes.

The overriding issue when recording environments for later listening and analysis is how memory and subjectivity might interfere with our interpretation of the experience. Interpretivists contend that it is the experiential moment that is important (Rabinow & Sullivan, 1988), and a recording can only ever be an indication of what sounds were in a space at a given time. It does not document how sound is perceived by the listener, but what it can do is document what impact a sound might have on other sounds. The recording technology became an extension of my listening experience (Ihde, 2007), but it was not the only process by which I was documenting the space. In recording this space, it was necessary to step back from the technology and instead pay attention to the entire sensory moment. The experience of listening and documenting became a holistic one, whereby the sounds, sights and smells were examined in context to each other. This means interpreting in what way each sensory experience impacts on others or how one sound might evolve in relation to other sounds. The following sections offer a detailed description of sites of listening exploring the links between the technological infrastructures of renewable energies, their potential impact on natural landscapes and their influence on local social, economic and cultural practices.

Listening to the Icelandic Soundscape

In 2015, I travelled to Iceland as part of a field recording group to document this varied and beautiful landscape and the sounds it produces. When documenting the soundscape, it was difficult at first to remove myself from the concern of recording noise[3]-free environments. However, on day three, I returned to the meditative practice of listening without judgement, an approach developed by Oliveros (2005). This argues for a deeper attention to the sonic environment. When one listens deeply, they discover not only the nuances of sound in space but also the ability to reflect on the meaning of these sounds in their everyday lives. I adopted a holistic approach to listening, analysing the meaning of all sounds in the environment, man-made or natural, reflecting on the relationship Icelandic people have with this landscape, and how the land is used.

During the field trips, an immense variety of sounds were recorded from birds within forests and marshes, to the gurgling, hissing and bubbling of sulphur pools, the explosions of steam and water from geysers, and the sounds of melting icebergs recorded underwater. In addition, we were recording during the summer period, where the sun never really sets; this meant we would record during the night and sometimes into the morning the very different wildlife that occupies various sonic terrains within the soundscape. Every day we would drive to a new space, set up our equipment and hope to document something different. Then in the evening, we analysed the day's recordings and shared stories about previous sonic experiences. One of the most interesting and strange animal sounds I heard on this trip was that of the common snipe; when it flaps its wings the sound is almost mechanical.

Renewable Energy Technologies: The Sound of Nature Harnessed

In recent years, Iceland's major energy source has moved towards hydroelectric power. There are a number of large powerful rivers, and these have been harnessed to feed the energy requirements of this relatively small and isolated country. This has meant that they have become less reliant on the importing of coal, oil or gas. In comparison to oil or gas production, the ecological impact is minor. On one of the field recording days when the group spent time inside one of these stations, it was this listening experience which began a deeper analysis of the ways in which listening might offer insights into potential ecological and social impacts.

Inside the station were five levels, with four below ground. At each level, the sound became louder and deeper in resonance, with the sound of turbines resonating and vibrating through the space and through our bodies. On the lowest floor, where the river was directed through the plant, the sound was at its most intense and pressed against the ear and the body. I used mainly binaural and contact microphones to document the resonances of this space. After only an hour of recording, I began to feel nauseous and was forced to leave the building. I was not alone in experiencing this embodied effect. Several others left, stating they didn't feel well and needed to get away from the continuous rumbling and whining of the many machines and turbines. Outside the station, the sounds were faint, but beneath our feet, the wave propagation produced by the turbines was travelling through the land and the river (Figs. 9.1 and 9.2).

On exiting the plant, it became impossible to ignore the potential of these resonating sounds to impact on subterranean or underwater ecosystems, the sounds radiated throughout this structure of concrete and metal. Low-frequency sounds have the potential to travel through objects and surfaces (Howard & Angus, 2009) and are known to cause physical reactions (Stocker, 2002). When outside the station I decided to place a

Fig. 9.1 Inside the hydroelectric power plant. (Photo by L. O Keeffe (author))

Fig. 9.2 Recording the river outside the power plant. (Photo by L. O Keeffe (author))

hydrophone in the river, the sound recorded was a constant repetitive low rumbling sound. A more effective form of documenting these sounds is required, over longer periods of time, to create an accurate analysis of these sounds and their potential impact on the wildlife of the river.

Researchers have suggested that oceanic and river environments are acoustically under threat. Helmreich (2011) and Stocker (2002) argue that oil drilling, factory fishing and pleasure boating produce anthropogenic noise, sounds caused by humans (Pijanowski et al., 2011), which, they counter, must affect marine life. It is well understood that the sounds produced by fish have a proven 'signal function' (Ladich 2015) necessary to their survival. Any sound, which masks or disrupts their ability to communicate, could damage an ecosystem. To date, there has been little investment in ecological studies or programmes to determine scientifically what the side effects may be of hydroelectric turbines. There is some research to indicate that hydrokinetic energy turbine sounds can have an impact on certain species of fish, with the strongest recorded impacts happening when sounds reached an average SPL level of "163 dB re 1 μPa for 24 h" (Schramm, Bevelhimer, & Scherelis, 2017, p. 1). Fish are extremely sensitive to low-frequency vibrations, usually associating those sounds to predators. This means they will swim away from that sound. This has the potential to create an "avoidance response" (BIAS, 2013) which could "affect normal behaviour related to movements, feeding, and reproduction" (Schramm et al., 2017, p. 1) resulting in significant "ecological consequences". In part, one could argue that because the ecosystem and soundscape of oceans and rivers are invisible and inaudible to humans, it has been easy to ignore the impact of certain technologies on the underwater soundscape.

While on a field recording trip to Iceland in 2015 one key sound acted as an intrusion on our ability to document the natural soundscape, which was the sound of tourism—a growing soundscape of tour planes and boats, and coach loads of people visiting remote sites from glacial lakes to volcanic mountains—these sounds masked our ability to record unique environmental sounds. The obvious anthropogenic sounds of cars, boats, planes and human voices were clearly identifiable and have often been defined as problematic to our environment. Yet it was the soundscape of the hydroelectric power stations that affected me the most, with the constant low-frequency pulses moving through the rivers and subterranean spaces of the Icelandic landscape. Their potential to harm an ecosystem is perhaps just as lethal as cars, boats and planes, but as an environmentally friendly energy, it is difficult to make a case to remove them as they cause little to no visible pollution.

The Terra Alta Soundscape: July 2015

The emergence of the Listening as Practice method in Iceland led to the development of new soundmapping techniques, new modes of listening and note taking, and finally the creation of a series of audio video art works inspired by the trip. In July 2015, following the trip to Iceland, composer Tony Doyle and I travelled to Spain to record the vast wind farms of the Terra Alta region of Northern Spain. In 2016 we created an acoustic ecology residency located in the village of La Fatarella and selected two artists to attend.

The Terra Alta area of northern Spain is a vast mountainous region, with summer temperatures reaching as high as 38 °C, leading to parched

Fig. 9.3 Wind turbines Terra Alta Mountains 2016. (Photo by L. O Keeffe (author))

landscapes, dry riverbeds, brittle fallen leaves and branches piling up on the ground. The winter temperatures reach well below 0 °C, with snow and freezing rains from October to March. Its agriculture is concentrated towards vineyards, almonds, cherries and olives, with a number of co-operatives producing jams, honey, olive oil and fruit liqueurs. I began listening to this region in 2008, and from 2014 onwards, there was a noticeable change in the natural and socio-cultural soundscape, which can in some ways be linked to the introduction of a new technological infrastructure, the wind farms (Fig. 9.3).

Listening to the Mechanical and the Natural

Composer Tony Doyle and I spent a month field recording both audio and video in the Terra Alta region. It was during this first Spanish field trip that the Listening as Practice method began to be refined. The method

Fig. 9.4 Recording the turbines and listening to crickets. (Photos by L. O Keeffe (author))

involved focusing on environmental sounds and socio-cultural sounds, including the documentation of stories, conversations and festivals. In addition, we were examining the potential of representing these findings in creative contexts (Fig. 9.4).

Most of the recording took place in the Serra de la Fatarella mountain range, a site of historical importance as it has many stone constructed dugouts created during the Spanish civil war. There are a number of walks that snake around the mountain allowing easy access to remote locations; during the day, the sounds of various vehicles, the boom of planes overhead and occasionally the sound of a tractor could be heard from all around. Embedded within this region are a significant number of wind turbines (see images), and the turning of the blades and engine within the base is heard clearly at a distance of approximately 100 metres. The sounds of this space including the wind turbines were recorded with a variety of microphones, from a boom microphone to a contact microphone and using the stereo inbuilt microphone of a zoom recorder. The process of recording the turbines involved moving outwards by degrees, focusing on how the sound changed as it was diffused by trees, bushes and wind, as well as the change in the physical landscape. What was very noticeable during these recordings was the soundscape of crickets around the turbines. When soundwalking through the mountains, as one approaches crickets, there is a sudden drop in sound as they respond to what they sense is a predator. It takes time sitting and not moving for the cricket soundscape to re-emerge. When walking around and near the turbines, even when setting up equipment, the crickets never stopped chirruping; it was as if they couldn't hear our sounds above the sound of the turbines.

In addition to recording the natural environment, a number of informal interviews took place which included recording the sounds of daily life in the village, documenting both the circadian (linked to farming practices) and the technological rhythms of the space dictated by other types of industry. A number of residents now commute to work in larger towns and cities. A decrease in local employment, including a reduction in farming work, has meant more of the community leave to work in cities, a statistic evidenced in the wider rural community in Spain and Europe.[4] The appearance of wind turbines in the region has meant, for some, a boost in finances; farmers receive a subsidy for every turbine on their land. This means there is less need to farm as much with this additional income. This has the resulting impact of changing the soundscape of these rural spaces, which were intrinsically linked to the biological rhythms of nature. The geographies of space and the social world are now merged with the social geographies of the technological world (Castells, Fernández-Ardèvol, Qiu, & Sey, 2009); this is increasingly true in remote areas with the emergence of renewable technological infrastructures. The countryside is now being shaped by the emergence of machinic rhythms within the circadian soundscape of natural and rural spaces. This can be seen as a mirroring of the changing soundscapes of cities since pre-industrial time, where certain aspects of the circadian rhythm of rural spaces still existed in city spaces (Lefebvre, 1992), but dissolved with the influx of mechanical technology (Fig. 9.5).

Fig. 9.5 Hybrid soundscape I–IV gallery exhibit for the *Sounds Like Her* exhibition, 2017–2020. (Photos by L. O Keeffe (author))

La Fatarella 2016: The Acoustic Ecology Residency

As a result of the initial 2015 field trip, Doyle and I decided to develop an acoustic ecology residency in 2016. We would select a small number of sound recordists and artists to locate themselves in this region in order to explore this changing soundscape. Additionally, in 2017, the artwork *Hybrid Soundscapes* and the paper "Listening to Ecological Interference: Renewable Technologies and Their Soundscapes" were presented at the Balance-UnBalance conference (L. O Keeffe, 2017a, b).

In August 2016, Robin Parmar and Matt Green were selected to participate in a 10-day acoustic ecology residency in La Fatarella. The residency included a number of excursions to different wind farms, the Ebro D'elta nature reserve and visits to a number of towns and cities. Parmar, Green and I also participated in a concert organised by SIRGA, a sound art organisation based in Catalonia, and presented our field recording experiences in the form of sonic art performances and discussions. In continuing with the Listening as Practice method, this field trip began with broader approach to examining the soundscape. It included more in-depth interviews with residents, documenting cultural events and activities in the region, and increased field recordings of the natural soundscapes, with a focus on documenting both audio and visual information. Additionally, I began to document my subjective experience of this area through a series of auto-ethnographic soundwalks; auto-ethnographies take the form of connecting the personal to the "cultural, social, and political" (Davis and Ellis 2013). Taking a holistic approach to listening means paying attention to the entire sonic environment, as the sounds of a place, both natural and man-made, play a large part in defining a space culturally, economically, socially and geographically (Fig. 9.6).

Listening became an integral part of noting differences in the environment; the drop in numbers attending festivals, the decreased farming sounds, the low numbers of children playing on the streets and the steady sound of the turbines whirring in the mountains. Additionally, a number of anecdotal stories emerged concerning the drop in bird sounds, particularly the sounds of swallows, and a decreased presence of wild boars and other small mammals in the woods. Locals were noticing small changes in the local ecology, with no immediate understanding as to why this was happening. There is little research on the impact of wind farms on wildlife; however, a recent study by the UK Royal Society for the Protection of Birds found that wind farms, while essential, if poorly sited, cause harm to

Fig. 9.6 Matt performing in Flix and the *Diablo Festival Tarragonna*. (Photos by L. O Keeffe (author))

bird populations from "disturbance, habitat loss (both direct and/or indirect) and collision" (RSPB, 2017). They have argued that it is necessary to examine what impact, if any, new technologies might have on the natural environment. Although the wind farms have come to provide a much needed source of income to these rural communities, there is evidence that the impact of this radically different economy is changing the face of these communities and perhaps affecting the natural ecology. Most of the farmlands in La Fatarella are passed from parent to child, providing a continuation of land practices, employment, community and cultural activities linked to this economy. La Fatarella has witnessed a population decline since the 1990s, from 1400 to 1004 in 2016.[5] This drop is seen as a real concern for the future of La Fatarella—without a young population, schools would close down, there would be little need to develop infrastructural projects in the village and property prices would drop due to a lack of buyers. Most properties in La Fatarella, like other small villages in the region, were bought and sold by families and locals. With the emergence of wind energy, there is no need for intensive farming, which might lead to disengagement with the land. Evidence shows that the younger population are leaving the region for larger towns and cities for employment and education opportunities (Delle Femmine, 2017), leading to an eventual silencing of these ancient rural communities.

Following the second field trip to Spain, I undertook a research trip to Beijing and to Barrow-in-Furness in the UK, and both trips allowed for a further development of the Listening as Practice method. In October 2017, this collection of material was presented as a body of artwork at the

Sounds Like Her exhibition, curated by Christine Eyene, at the Nottingham Art Exchange Gallery. The exhibition was an attempt to represent the immersive experience of data collected and analysed from each sonic environment since 2015.

Conclusion

The development of the Listening as Practice approach allows researchers to make audible and visible the impact of new technological infrastructures on different biological and social systems. Sound as a phenomenological experience is non-linear, exists in multiple spaces simultaneously *and* happens over time. Its influence is not localised to its source, and in the process of examining the effect of a soundscape, one must look at all the spaces for which it might have an impact. By including an ecological perspective with a social analysis of the interaction and impact these systems have on the living conditions of humans and other species, we highlight the deep and necessary links needed to create a balanced ecosystem. Additionally, the Listening as Practice method attempts to engage a broader audience with community-engaged art processes like soundwalking and soundmapping projects,[6] as well as finding new ways to present findings through performance and exhibition. This means that the research findings are not limited to academic circles and have the potential to shape a communities engagement with a new technological infrastructure.

The chapter does not suggest discarding renewable energy technologies but applying a new level of interrogation into any potential impacts it might have on an ecosystem, regardless of its overall benefits. The systemic use of technology in everyday life has definitively altered our relationship to space and ideas, and technological determinists would argue that this is of course the natural process of a ubiquitous technological world. However, this has the potential to promote a passive attitude to any technological change. The development of technological systems or objects is neither linear nor inevitable (Wajcman & MacKenzie 1999). Technologies are embedded with the social and political interests of a culture, and the design of technologies is a constant and open process, dependent on the needs and circumstances of society (Klein & Kleinman 2002). In listening to the soundscape of a technological system, one can become alert to the potential impact on a biological and social system. Using Listening as Practice allowed for the collection of materials, which highlighted systemic changes in the social and natural environment. This meant standing back from the larger ecological and environmental goals of renewable technologies and focusing on macro technological soundscapes and the hegemonies that control this new auditory environment. Combining techniques such as soundmapping, drawing, recording and listening allows for a more in-depth collection of data, which reveals interconnections between diverse spaces, objects, technologies and communities. It is at the point of hearing the whole that one understands how a shift in the soundscape can have a ripple effect beyond the source of that sound.

Notes

1. See https://www.conserve-energy-future.com/pollutiontypes.php
2. See http://anjakanngieser.com/about/
3. In this instance, I use the term noise to indicate anthropogenic sounds, such as cars, planes and people.
4. http://ec.europa.eu/eurostat/statisticsexplained/index.php/Agricultural_census_in_Spain#Main_statistical_findings
5. Population statistics provided by https://www.citypopulation.de/php/spain-cataluna.php?cityid=43056
6. See work undertaken in Singapore in 2016: http://lindaokeeffe.com/blog/#/blog/composing-the-singapore-soundscape-at-nus/

References

Barclay, L. (2017). Listening to Communities and Environments. *Contemporary Music Review, 36*(3), 143–158. https://doi.org/10.1080/07494467.2017.1395140

BIAS. (2013, November 4). *Fish and Sound*. Retrieved June 1, 2017, from https://biasproject.wordpress.com/news-from-the-ocean/fish-and-sound/

Carson, R. (2000). *Silent Spring* (New Ed edition). London: Penguin Classics.

Castells, M., Fernández-Ardèvol, M., Qiu, J. L., & Sey, A. (2009). *Mobile Communication and Society: A Global Perspective*. Cambridge, MA: The MIT Press.

Davis, C., & Ellis, C. (2013). Emergent Methods in Autoethnographic Research Autoethnographic Narrative and the Multiethnographic Turn. In S. N. Hesse-Biber & P. Leavy (Eds.), *Handbook of Emergent Methods* (pp. 283–302). New York: Guilford Publications.

de Certeau, M. (1988). *The Practice of Everyday Life*. Berkeley, CA: University of California Press.

Delle Femmine, L. (2017, November 6). Room in the Middle: The Africans Repopulating Spain's Dying Villages|World News|The Guardian. *The*

Guardian. Retrieved from https://www.theguardian.com/world/2017/jun/11/room-in-the-middle-the-africans-repopulating-spains-dying-villages

Epstein, M. (2009). *Growing an Interdisciplinary Hybrid: The Case of Acoustic Ecology*. Retrieved December 29, 2009, from http://www.ucalgary.ca/hic/issues/vol3/9

Helmreich, S. (2011). Underwater Music: Tuning Composition to the Sounds of Science – Stefan Helmreich. In T. Pinch & K. Bijsterveld (Eds.), *The Oxford Handbook of Sound Studies*. Oxford, MS: Oxford University Press.

Howard, D. M., & Angus, J. (2009). *Acoustics and Psychoacoustics*. Amsterdam/Boston/London: Focal Press.

Ihde, D. (2007). *Listening and Voice: Phenomenologies of Sound*. Albany, NY: SUNY Press.

Klein, H. K., & Kleinman, D. L. (2002). The Social Construction of Technology: Structural Considerations. *Science, Technology & Human Values, 27*(1), 28–52. Retrieved from http://sth.sagepub.com/content/27/1/28.short

Ladich, F. (2015). *Sound Communication in Fishes*. New York/Berlin/Heidelberg: Springer.

Lefebvre, H. (1992). *Rhythmanalysis: Space, Time and Everyday Life*. London/New York: Continuum.

O Keeffe, L. (2015a, April 20). *Workshopping the Derry Soundscape: Mobile Technologies as Creative Tools for Third Age Adults*. Retrieved September 26, 2017, from https://soundstudiesblog.com/2015/04/20/mobile_technologies_for_third_age_adults/

O Keeffe, L. (2015b). Memories of Sound: Socioeconomic, Community and Cultural Soundscapes of Smithfield, Dublin from the 1950s. In L. Back & M. Bull (Eds.), *Auditory Culture Reader* (2nd ed., pp. 217–228). Oxford, UK: Berg.

O Keeffe, L. (2017a). Ecological Interference: Hybrid Soundscapes. In *A Sense of Place*. Plymouth, UK: Plymouth University.

O Keeffe, L. (2017b). Listening to Ecological Interference: Renewable Technologies and Their Soundscapes. In *A Sense of Place*. Plymouth, UK: Plymouth University.

Oliveros, P. (2005). *Deep Listening: A Composer's Sound Practice*. New York: iUniverse, Inc.

Pijanowski, B. C., Villanueva-Rivera, L. J., Dumyahn, S. L., Farina, A., Krause, B. L., Napoletano, B. M., et al. (2011). Soundscape Ecology: The Science of Sound in the Landscape. *Bioscience, 61*(3), 203–216. https://doi.org/10.1525/bio.2011.61.3.6

Rabinow, P., & Sullivan, W. M. (1988). *Interpretive Social Science: A Second Look*. Berkeley, CA: University of California Press.

RSPB. (2017). *Wind Farms*. Retrieved June 19, 2017, from https://www.rspb.org.uk/our-work/our-positions-and-campaigns/positions/windfarms/index.aspx

Schramm, M. P., Bevelhimer, M., & Scherelis, C. (2017). Effects of Hydrokinetic Turbine Sound on the Behavior of Four Species of Fish Within an Experimental Mesocosm. *Fisheries Research, 190*, 1–14. https://doi.org/10.1016/j.fishres.2017.01.012

Stocker, M. (2002). Ocean Bio-Acoustics and Noise Pollution: Fish, Mollusks and other Sea Animals' Use of Sound, and the Impact of Anthropogenic Noise in the Marine Acoustic Environment. *Soundscape*, *3*(2), 1–26.

Wajcman, J., & MacKenzie, D. (1999). Introductory Essay: The Social Shaping of Technology. In D. MacKenzie & J. Wajcman (Eds.), *The Social Shaping of Technology* (2nd ed.). Buckingham, UK: Open University Press.

Whitehouse, A. (2015). Listening to Birds in the Anthropocene: The Anxious Semiotics of Sound in a Human-Dominated World. *Environmental Humanities, 6*, 53–71. Retrieved from https://doaj.org

SECTION III

Media and Society

CHAPTER 10

The Uncanny Soundscapes of the Palestinian Exile: Rethinking Technics, Memory, and Sound

Özgün Eylül İşcen

First, we see a few lines of Arabic script on the screen, which is superimposed onto a blurry close-up shot in the background. Soon thereafter, we start hearing a taped conversation in Arabic between Mona Hatoum and her mother recorded in Lebanon, which is kept indecipherable for the ears of the foreign audience. The still images slowly fade into similar close-ups of the mother's naked body in the shower shot by Hatoum. In the meantime, more lines of the letters, sent by Hatoum's mother to her daughter, keep floating over the screen and obscure the close-ups of the mother's body in the background. Then, we hear Hatoum's voice slowly reading the letters translated into English. The conversation between Hatoum and her mother, through their voices and letters, develops around the mother speaking about war, family relations, and sexuality, including her husband's objections to Hatoum's recording of her mother's naked body. This intimate video about a mother–daughter relationship is one of the most well-known works of the Lebanon-born Palestinian artist Mona Hatoum called *Measures of Distance* (1988).[1]

Ö. E. İşcen (✉)
Duke University, Durham, NC, USA
e-mail: ozgun.eylul.iscen@duke.edu

M. Droumeva, R. Jordan (eds.), *Sound, Media, Ecology*,
Palgrave Studies in Audio-Visual Culture,
https://doi.org/10.1007/978-3-030-16569-7_10

Mona Hatoum was born in 1952 to a Palestinian family living in Beirut in exile after their displacement from Palestine. She grew up in Beirut, but embraced London as a second home after a visit in 1975, just before the civil war broke out in Lebanon, and continued for the next fifteen years. As a Palestinian woman living in the West as an exile, her works mostly deal with the issues of longing, witnessing the violence of war from a distance, and finding a voice for the ongoing struggle of the Palestinian people. In her early works, Hatoum carried out several performances, in which she used her body, breath, and voice in opposition to the mainstream representations of the wars happening in her homeland (such as soundscapes of newsreels). In these works, Hatoum did not only seek to challenge the various forms of misrepresentation and silence/silencing about the Palestinian struggle but also attempt to remember and speak for herself, and for her people.

In this sense, Hatoum's *Measures of Distance* is an example of what Hamid Naficy (2001) calls "accented cinema", which is defined as an aesthetic response to the experience of displacement through exile, migration, or diaspora. Accented cinema derives its accent from its usually multilingual nature and its mediation of different cultural and symbolic regimes. Not surprisingly, Naficy's accented cinema relies on acoustic metaphors related to voice, speaking, and communication. In Hatoum's case, it takes the form of "epistolary style", which refers to an autobiographical response to the experience of displacement (Naficy, 2001, pp. 118–120). For Naficy, the epistolary media (like letters, albums, diaries, and video/audio recorders) play a constitutive role in the lifeworld of displaced people. It is with them that they construct their past and current identity.

The acousticity of accented epistolary style derives from the specific inscription of sound and voice, such as the contrast between the rapidly spoken Arabic and the slower fragile English voice of Hatoum. In his reference to Walter J. Ong,[2] Naficy underlines that sight distances or isolates the seers from the seen, whereas sound unifies the maker or sender of the sound with its recipient (2001, p. 120). As Laura U. Marks (2000) puts it, Hatoum's voice reading the letters, in addition to the haptic visuality[3] of the layered video screen, calls upon sense memories to restore the history that has become externalized and fossilized in an object, such as letters (Marks, 2000, pp. 154–155). As apparent in the still image from *Measures of Distance* depicted in (Fig. 10.1), the layers of the video constitute a thick surface that almost feels tactile. This tactility of voice and haptic image directly addresses the body of the audience and destabilizes any

Fig. 10.1 A still from *Measures of Distance* (Hatoum, 1988). http://www.tate.org.uk/art/artworks/hatoum-measures-of-distance-t07538

one-sided experience or meaning of those recorded objects of memory. In *Measures of Distance*, the intimacy of the taped content and voices offer a different entry into the Palestinian context, beyond its fetishized media representations that underscore either trauma or nostalgia.

Inspired by Hatoum's work, this chapter investigates the act of recording and composing sounds that survive over time and distance within an exilic condition. Like Hatoum's sound diary that is carried over Vancouver on a tape, these recordings survive as materials for making sense of one's life and for telling one's story. Since there is no return home in the exilic setting, one's past and present are mediated by the objects of recollection. In this sense, the exilic condition creates a provoking example for Bernard Stiegler's theory of technologically mediated memory, especially his concept of "tertiary memory", which refers to a memory that is not experienced live but only made accessible through its technological mediation. Furthermore, Stiegler complicates the operation of "tertiary memory" by asserting his concerns about the industrialization of memory, which underscores the transition from individual exteriorization of memory to mass-consumed, industrial temporal objects (like radio, television, cinema) that organize consciousness for certain ends under the current capitalist system (2011).

Following upon Stiegler's work, this chapter asks how technologically mediated soundscapes shape one's negotiation of the conditions of exile. In the case of a Palestinian artist like Hatoum, tertiary memory is the only way left to be bound with her homeland, while it is the only way to encounter her exilic condition for the foreign audience, like the ones in Vancouver. For Palestinians, the temporal/spatial disjunction entails physical, mental, metaphysical, moral, and various non-territorial geographies (Tawil-Souri, 2014, p. 174). As Helga Tawil-Souri describes:

> Being Palestinian is having to live with, (re)negotiate, challenge, and resist various mechanisms and power-struggles over movement and sedentariness—checkpoints, refugee camps and the more abstract "spaces" of occupation, diaspora and exile, of laissez-passer documents rather than "real" passports, of travel permits, and so on. Dealing with various time-space processes that weave disparity, disconnection, fragmentation, and un-synchronization—determined by another state—into the fabric of life is an experience shared by all Palestinians. (2011, p. 21)

Thinking of Stiegler's insights into mediated memory through the "accented cinema" of artists like Hatoum raises significant questions about the role of sound in shaping the experience of exile, which resists a linear chronology or geography. I will demonstrate how the field of acoustic ecology shares with Stiegler an important critique of the industrialization of memory, which provides necessary tools to address the role of mediated sound in cinematic manifestations of the exilic condition. At the same time, I will show the limits inherent in acoustic ecology's prescriptive emphasis on balanced "ecology" by reimagining issues pertaining to sound, identity, and exile around what Iain Foreman calls "acoustic ontology" (2011, p. 266), better acknowledging the experience of "uncanny soundscapes". By drawing upon Jean-Luc Nancy's ontological conception of space, Foreman redefines acoustic community as one of absences, fragments, and lacunae, which is marked by its inability to have a substantive identity rather than sharing a balanced commonality (2011, pp. 268–269). Indeed, the context of Palestinians living through ongoing violence, injustice, and silencing, whether in exile or diaspora, inside Israel or within the Occupied Territories, underscores the contradictions of identity and collective memory as a point of departure. In this regard, I examine the practice of soundscape composition, the main creative offshoot of the field of acoustic ecology, as the basis for this reframing, ultimately suggesting that this creative practice provides a template for grasping how media practice is being used to articulate the subjectivity of global patterns of displacement.

Technics and Memory in Stiegler

In his multivolume work *Technics and Time*, Bernard Stiegler reflects on the relationship between human and technical objects, or what he calls technics. In the first volume of the series, *The Fault of Epimetheus* (1998), Stiegler challenges the modern interpretation of technics in terms of a means–ends relationship[4] and puts forward the idea that technology is an ontological constituent of the human being. Drawing upon Martin Heidegger, Gilbert Simondon, André Leroi-Gourhan, and Jacques Derrida, Stiegler claims that technics becomes a vehicle for humans to become human since they are constituted through their coupling with technics. Following upon his emphasis on the default origin of human, Stiegler argues that technics is prosthetic in terms of a constitution of human rather than a mere extension of it (p. 152). With this constitutive role given to prosthesis as technics, Stiegler transforms our understanding of externalization along the Derridean path he builds on.

For tackling the close relationship between the notion of externalization and history, Stiegler works through Heidegger and Derrida closely. In Heidegger, Being (Dasein) does not have an origin or essence. A Being is constituted by the "already-there", which is the past,[5] and the sense of "not-yet", which is the future (1998, p. 5). Therefore, for Heidegger, Being cannot be grasped except by taking the notion of time into consideration. Stiegler is interested in how the temporality itself is constituted in terms of technics and claims that the technological time—the inauthentic time in Heidegger's terms—is constitutive of Being (1998, p. 215). For instance, Heidegger interprets the clock's "fixing" time from a negative perspective, whereas, for Stiegler, "fixing" time does not mean to constrain but to establish it (1998, p. 220). The possibility of calculation forms the tradition itself for which the instruments open up a publicity, datability, or significability.

Stiegler concludes that Being is always already "inauthentic", since it is a prosthesis from the very beginning, and Being's access to its past, and thus its future as such, is prosthetic as well (1998, pp. 233–235). Stiegler translates this constitutive role given to the technological object into the

Derridean terms by saying that Being is constituted by the "différance" through the process of exteriorization (pp. 134–142). Since there is nothing given before its externalization, Being cannot be thought separately from the tools or artificial supplements. More importantly, every epoch is determined along the technical conditioning of memory by the process of grammatization, which refers to a process of discretization of production and experience (e.g. binary digits and algorithms of the digital era). As Stiegler emphasizes, "every epoch is characterized by the technical conditions of actual access to the already-there that constitutes it as an epoch, as both suspension and continuation, and that harbor its particular possibilities of 'differentiation' and 'individuation'" (1998, p. 236). In the second volume of the series *Disorientation*, Stiegler (2008) continues to address the relationship between the technical object and time-consciousness through his reworking of Edmund Husserl. From Husserl, Stiegler takes the idea of a temporal object, such as melody, that is not simply in time, but constitutes itself temporally (2008, pp. 198–200). For instance, when one listens to a melody, the object we hear is in a constant flow of appearing and vanishing, and thus it occurs in the interplay of past and future rather than an isolated moment. According to Husserl, the perception of tones as a melody is constituted by the fading away tones that do not vanish totally from our consciousness (but keep fading away) and the sense of continuity with an expectation that something would arrive next (Stiegler, 2011, p. 14). Therefore, within this flow of melody, there are three components: the actual live experience of a tone (primary impression), primary memory of the tone fading away that is not present to our perception anymore (primary retention), and the anticipation of further tones in the melody (protention) (Stiegler, 2008, pp. 219–233). Stiegler is interested in the distinction Husserl makes between the primary retention and secondary retention, which refers to a memory of a past moment like a melody heard a day ago. The retention of previous tones belongs to the very act of perception since there is no perception of the melody without primary retention (Stiegler, 2011, p. 14). Hence, Husserl differentiates this type of primary memory from what he thinks of as secondary retention or memory, which is a repetition of a past moment in the imagination. As Stiegler emphasizes, for Husserl, the crucial point is that primary retention belongs to the act of perception, whereas secondary memory belongs to the one of imagination. However, Stiegler challenges the sharp opposition between primary and secondary memory in Husserl by questioning the purity or primacy of primary retention (2008, pp. 223–226). For Stiegler, it is not the same thing we hear if we listen to a melody twice or more. Each time, our ears hear differently, because the previous listening has an impact on our ears (2008, pp. 53–57). This is where the tertiary retention (media and culture) comes to the foreground, since it is constitutive of both primary (sense-perception) and secondary (memory) retentions.

In the third volume of the series *Cinematic Time and The Question of Malaise*, Stiegler (2011) draws upon cinema as a further example of Husserl's temporal object. For Stiegler, cinema is a temporal object due to the photo-phonographic coincidence of past and present, and the coincidence between the film flux and the flux of consciousness of the film's spectator (2011, pp. 11–12). This is how our consciousness adapts to the time of industrial temporal objects such as cinema, which is controlled by the contemporary profit-driven and standardized media industries and information technologies (Stiegler, 2011, pp. 17–18). Cinema as a tertiary memory partakes in the history of exteriorization and exploitation of our consciousness. Following the Kantian schematism, and its critical appropriation in Theodor W. Adorno and Max Horkheimer's critique of "culture industry", Stiegler voices his concerns about the industrialization of consciousness and the malaise from which the current global culture suffers as a consequence.

In *Symbolic Misery 1* (2014), Stiegler offers a detailed analysis of Alain Resnais' film *Same Old Song* [*On Connaît la Chanson*] (1997) to underline how tertiary memory constructs one's subjectivity. Stiegler reinterprets Simondon's notion of the preindividual, which is the reserve of potentials for future individuations, in terms of "already-there" (2014). In other words, Stiegler resituates the preindividual as tertiary memory, and thus claims that the collectively shared cultural artifacts like songs and films shape our singular individuations. In the film, the characters sing songs of French popular culture in various scenes. More interestingly, the characters mostly mime or sing along the songs (rather than lip-synching as a song's singer), like we do in our everyday life. In Stiegler's words, the characters "are, LIKE US, inhabited by the songs that animate them" (2014, p. 25). For Stiegler, these songs fashion the characters or the audience whose own memories do those songs evoke. In the critical moments, the film characters take on the character or mood of the song they are singing; or in Heideggerian terms, inheriting "we" (the past as "already-there" that they haven't lived per se). They know those songs without deliberately listening to them, or citing those songs without even realizing it. For Stiegler, this "every" of every one of us is an impersonal "one" rather than collective "we". The contemporary culture industries constitute a massive synchronization machine that flattens the depth of time and individuation (Stiegler, 2014). Nonetheless, Stiegler's insistence on the coupling of human and technics encourages him to find a way out of this symbolic misery without turning against the technology itself.

Encounters Between Stiegler and Soundscape Studies

Stiegler's critique of industrialization of memory (and, thus, of consciousness) under the current capitalist system is a common ground for the various perspectives that can be brought together under the title of soundscape studies. The World Soundscape Project (WSP), where the roots of soundscape studies come from, aimed at addressing the changing soundscapes and acoustic communities starting in the late 1960s due to a period of rapid industrialization and urbanization. The WSP, and its related fields of acoustic communication (Truax, [1984] 2001), and acoustic ecology highlighted the changing dynamics of social mediation triggered not only by changing soundscapes but also by the developing technologies, from machinery in factories to recording devices in studios. In *The Tuning of the World* (1977), R. Murray Schafer, who set up the WSP in the late 1960s, underlines that the sound of a particular locality (its keynotes, sound signals, and soundmarks) could evoke personal memories, and construct a collective identity to the extent that settlements can be recognized by their soundscapes. Since the industrial revolution, however, an ever-increasing number of soundscapes have either disappeared or submerged into the cloud of homogenized anonymous noise that the urban centers are embedded within. In Stiegler's terms (as he refers to Simondon), Schafer was concerned about this loss of individuation in the context of changing soundscapes at that period.

Furthermore, the WSP was cautious about the prevalence of technologically mediated soundscapes, sounds that are removed from their original context and played back in an environment other than the one in which they were created (Truax, 2001, pp. 134–135). This is what Schafer calls the "schizophonic" effect, referring to the split between the environment that the listener is in and the world that is projected by the electronically reproduced and transmitted sounds (1977, pp. 90–91). Parallel to Stiegler's emphasis, the schizophonic effect underscores how technological mediation plays a significant role in shaping psychic and collective individuation in a given epoch. As Hildegard Westerkamp (1988) notes, the schizophonic effect speaks to the pervasiveness of the imposition of electroacoustically mediated sounds and voices in our everyday life, from workplaces to shopping malls (p. 27). Westerkamp offers an analysis of music-as-environment (e.g. Muzak[6]), which is designed not to be listened to but to accompany other, mostly commercial, activities. Similar to Stiegler's concerns over the industrialization of consciousness (and the proletarianization of consumers), such environments do not only force us to participate in the activity of commodity exchange but also rob our desires to listen and make sounds (Westerkamp, 1988, p. 2).[7] As Westerkamp emphasizes, there is a close tie between the regressive listening (in Adorno's terms) and silencing of voice (1998, p. 100). In Stiegler's terms, we can say that the repression of listening and making sound leads to the symbolic misery, which does not only refer to the ways of how our consciousness is appropriated and exploited but also how our desires are manipulated and repressed:

> By symbolic misery I mean, therefore, the loss of individuation, which results from the loss of participation in the production of symbols. Symbols here are being as much the fruits of intellectual life (concepts, ideas, theorems, knowledge) as sensible life (arts, know-how, mores). And I believe that the present status of generalized loss of individuation can only lead to a symbolic collapse, or collapse of desire - in other words to the decomposition of the social as such: the total war. (Stiegler, 2014, p. 10)

In this regard, I argue that soundscape studies offers critical and creative practices that embrace such a gesture of intervening into the symbolic realm by reclaiming and manipulating its production. Practices like soundwalking, field recording, alternative radio programming, and soundscape composition can be seen as efforts toward changing our habitual, standardized ways of listening and sound-making, at both psychic and collective levels. For instance, soundscape composition has become a way of deconstructing our habitual listening and recontextualizing the "found" sounds for certain aesthetic and political ends. More importantly, it aims at enhancing our awareness of listening and sound-making, while triggering a desire for such experimentation and reinterpretation. In Stiegler's terms, soundscape composition invests in the composer as a participant in its own resingularization, who reworks the process of psychic and collective individuation through technical mediation.

Stiegler's triple model indicates an ecological perspective since it brings the processes of technical, psychic, and collective individuation together. The act of reworking of "already there" through a technologically mediated intervention can generate the modes of individuation that are not possible otherwise. Therefore, what makes soundscape composition unique is its tie to found sounds and acoustic environments ("already-there"), while repeating it with a difference. For instance, in her composition *Cool Drool* (1987), Hildegard Westerkamp reassembles the pieces of already existing popular music that we hear in our daily life in order to make sense of background music for consumption; "it goes as far as playing the entire pieces of music, which were composed by others, back to the listener." Similar to Stiegler's analysis of Resnais' *All Old Songs*, the audience is exposed to the complexity and banality of everyday songs that shape our patterns of consumption and communication.

The agenda and methods set by soundscape studies contribute to Stiegler's critique of industrialization of consciousness. More importantly, creative practices like soundscape composition underscore the significance of reworking experience through technological mediation for the task of intervention. However, their mutual critique does not fully address what I describe as the exilic condition in this chapter. While agreeing with their critique of loss of individuation, I argue that the exilic condition comes with more fragmented memory situations and a limited access to possible interventions. The exilic condition motivates a negotiation of one's current conditions as well as memories, especially if their migration is forced by political circumstances. Thus, the aesthetics and politics of exilic soundscapes encourage us to readdress any notion of "balance" to begin with.

The Uncanny Soundscapes in Basma Alsharif's *Deep Sleep*

This chapter is interested in the soundscapes that are fragmented from the very beginning, sonic memories of ruined soundscapes and displacement. In his article titled "Uncanny Soundscapes", Iain Foreman (2011) highlights the conceptual space shared between trauma and sound. For Foreman, both sound and trauma operate through unsettled temporality and spatiality that heighten the entangled relationship between present/past, presence/absence, and inside/outside. He addresses this complex space by asking "How is trauma remembered or unremembered through sound and listening? Do soundscapes bear witness to trauma? [...] Does listening, or do soundscapes, provide a response to trauma, or a space in which trauma can be presented?" (Foreman, 2011, p. 266).

My interest is not located in framing the complex memory situations of exile through the lens of trauma; however, Foreman's emphasis on the conceptual space between sound and trauma reconfigures the idea of ecologically balanced soundscape or community, in order to address the shattering experiences of exile, marked by dislocation and loss. For Foreman, "the spatiality of the ruins, the fragmentation of memory, and the aporias of testimony" (2011, p. 270) could only recognize the incompleteness of being, and being together, characterized by sharing and splitting at once. With his reference to Jean-Luc Nancy's notion of "inoperative community", Foreman redefines acoustic community as one of absences, fragments, and lacunae, which is marked by its inability to have a substantive identity rather than sharing a balanced commonality (2011, pp. 268–269). By drawing upon Nancy, Foreman argues that:

> [I]t is necessary to rethink soundscape composition in terms of acoustic ontology rather than acoustic ecology by arguing that, following Nancy's ontological approach in particular, a soundscape's being-in-the-world gravitates around the concepts of splitting and sharing (partage), both presupposing a relation to the outside, an exposure to exteriority rather than a logic of identity.[8] (Foreman, 2011, p. 266)

According to Foreman, the complexities of remembering and testifying through listening and composing generate an uncanny experience at the edge of presence and absence, of intimacy and distance. This is what Foreman calls an "uncanny soundscape". The effect of dislocation—the schizophonic effect in Schafer's term—brings the listener/sound-maker closer to the condition of exile, rather than phenomenologically grounding them in a determinate place and time. In his reference to Nancy, Foreman underlines that listening is not necessarily a metaphor for access to self, but the reality of this access: "a reality consequently indissociably 'mine' and 'other', 'singular' and 'plural', as much as it is 'material' and 'spiritual' and 'signifying' and 'a-signifying'" (as cited in Foreman, 2011, p. 269).[9] Instead of monumentalizing and reproducing a homogeneous identity or narrative, soundscape composition has the capacity to foreground the interruptive and incomplete nature of displaced memories. As Foreman argues, such a creative yet disruptive organization of soundscape makes memory and testimony possible, and, as such, enables a presentation of an impossible communication, testimony itself.

Against the background of the ongoing aggressive occupation of Israel and silent complicity of the Western and Arab countries, the issue of testimony is much more multi-situated for the Palestinians. The Palestinian memory harks back to a traumatic originary event (the Nakba), but, at the same time, it is constantly reworked, reactivated by new events, and rearticulated in new acts of memory (Saloul, 2012, p. 211). The Nakba—the Catastrophe—of 1948 marks the initial trauma when the state of Israel was created and the Palestinians were denied their land. Since then, Palestinians have been living in an ever-diminishing space, which underlies their exilic conditions of existence within or beyond the land occupied by Israel (Matar, 2011). The collective ontology of Palestinianness has been mostly shaped by this constant negotiation of the tensions between presence and absence, as well as mobility and immobility (Tawil-Souri, 2014).

Nonetheless, as Tawil-Souri argues, Palestinian cinema serves as a space to transgress Palestine's spatial, temporal, mental, and psychological entrapment. In other words, it unsettles the exclusivist logic of territorial-national imaginary by enacting a transgressive constitution of self, place, and nation:

> In the words of Hamid Dabashi, 'the overriding presence of an absence is at the creative core of Palestinian cinema, what has made it thematically in/coherent and aesthetically im/possible' (2006). (…) Space here can be largely negative: disconnecting, separating, differentiating, hierarchizing. But, space—and its absence—can equally be more open-ended, alternative, innovative, radical, messy, emotional and non-linear. It is not a matter of dichotomy but of contradictions in a true dialectical movement.[10,11] (Tawil-Souri, 2014, p. 184)

As a creative expression of such dialectical movement, in *Deep Sleep* (2014), Kuwait-born Palestinian Basma Alsharif[12] draws upon her exilic experience of nomadic life, belonging to multiple places and histories at once. As she is regularly restricted from traveling to Palestine, she appropriates a technique of autohypnosis for the purpose of bi-locating. *Deep Sleep* was recorded on 8 mm film and built on brainwaves that generate binaural beats. The video takes the audience on a journey through the soundscapes of different sites of modern ruins, in Athens, Malta and the Gaza Strip.[13] With *Deep Sleep*, Alsharif reconstructs historical, geographic, and cultural resonances across the Mediterranean civilizations. Influenced by her own familial and personal relationship to Palestine, Alsharif seeks to offer a sense of "bi-location" that transcends the limits of geographical borders, and plays with the thick layers of memory and history. As Alsharif puts it, her idea was to investigate how she could be in two places at once, so she could be in Gaza when she could not be. For this aesthetic end, in *Deep Sleep*, Alsharif works with binaural beats. A binaural beat is an auditory illusion perceived when two different pure-tone sine waves, both with frequencies lower than 1500 Hz, with less than a 40 Hz difference between them, are presented to a listener dichotically (one through each ear). In less technical terms, each ear is activated separately. By appropriating binaural beats, Alsharif develops an aesthetic strategy for triggering a dislocating effect—an uncanny feeling that is the fundamental aspect of the exilic condition.

In *Deep Sleep*, Alsharif sensorily diverts the audience to remotely engage them in Gaza, while triggering questions about how civilizations are established, endured, and destroyed (Fig. 10.2). The viewer is visually and aurally immersed in the ambiguous spaces between hypnosis, fantasy, and reality. Gaza is imagined and reimagined through the ruins of Malta and

Fig. 10.2 A still from *Deep Sleep* (Alsharif, 2014). https://vimeo.com/101257097

Greece, and the linear history is thrown into question, as the past, present, and future are never made certain. As Alsharif puts it herself:

> That film was all about my perspective, my body, and of moving through various sites in various states. The viewer was meant to occupy my position in watching the film, and to be reminded that it was not their own but that they could easily be a part of it. This was important, as it was a very direct attempt at addressing the end of civilization in Gaza: a place that is being wiped clean of its historical monuments. While in other sites (Malta, Greece) the historical monuments are a testament to great civilizations past, perhaps deterring us from addressing the crisis facing those cities today.[14]

What is striking about *Deep Sleep* is how Alsharif's visual and aural compositions operate both at an ontological level (closely tied to the experiential conditions of her nomadic life) and at the collective level (where her psyche meets with the already-there of the historical). For Palestinians, one's memory has been already fragmented, and the "already-there"/tertiary memory has been also destroyed or rendered inaccessible. In *Deep Sleep*, Alsharif makes and records sounds around the ancient ruins, while bringing these different historical sites into a dialogue. It is her attempt to relate to the "already-there" and demonstrate how this relationship is politically and economically controlled. In Stiegler's terms, Alsharif literally reclaims the control of her flux of consciousness, and as such, she triggers a resingularization of her story, of the places she travels through, and of the encounters she offers to the audience. More importantly, like Hatoum, Alsharif shows us how such a reclaim activates, if not derives from, an inherently political desire for addressing such a complex historical condition.

Conclusion

This chapter investigates the question of technics and memory in the context of mediated soundscapes of exile, through the dialogue between Stiegler and soundscape studies. On the one hand, Stiegler's tertiary memory embraces a more crucial role within exilic settings. The memory situations of exile complicate the dynamics of primary, secondary, and tertiary memories, and thus open up to a richer process of resingularization of "I" and "We", as a critical response to the loss of individuation in Stiegler's terms. On the other hand, soundscape studies shares some common ground with Stiegler's critique of industrialization of memory, and offer some aesthetic strategies, like soundscape composition, in order to trigger such resingularizations. According to Stiegler's model, the exilic condition can be interpreted as an interruption in the psychic and collective realms through which the dominant modes of symbolic production operate. Like Alsharif's *Deep Sleep*, Stiegler encourages us to intervene in the technical individuation—the production of tertiary memories—that mediate the relationship between psychic and collective individuation.

The political desires and aesthetic experimentations that are entangled with the Palestinian exilic conditions underline both Hatoum's and Alsharif's works. Both artists intervene in the realm of tertiary memory, shaping the predominant representations of the Palestinian struggle and Arab womanhood, which is also the reserve for future individuations. For instance, Hatoum's conversation with her mother about sexuality and war negotiates with various patriarchal discourses around these issues, while her field recording and editing make the survival and intimacy of that conversation possible. As Westerkamp emphasizes, it requires effort to learn how to listen, make sound, and find one's voice, especially as a member of the marginalized social groups, whether a woman or an immigrant in exile. In *Deep Sleep*, Alsharif draws upon her experience both at an ontological level, which is fashioned by her own exilic life, and at a collective level, which expands to a larger geography that is politically and economically weakened.

I would like to conclude this chapter by highlighting its relevance to contemporary forms of technological mediation, and digital and networked media. This chapter demonstrates how soundscape studies can contribute to current debates by offering historical and aesthetic perspectives on the issue of industrialization of consciousness, from the changing sensory capacities to the military-industry underpinnings of developing technologies. As both Stiegler and soundscape studies emphasize, it is important to develop aesthetic strategies to rework the hegemonic applications of contemporary technologies. One may question the significance of artistic interventions; however, artistic responses help us to understand the material conditions and dominant ideologies that those technologies operate through. Finally, the exilic condition brings an alternative aesthetic style and political desire into the picture and, as such, encourages us to historicize and contextualize the impacts of contemporary technologies in their multi-situatedness.

Notes

1. Mona Hatoum recorded the slides and conversations in Beirut in 1981. She produced *Measures of Distance* at the artist-run and Vancouver-based Western Front Art Centre in 1988 with the materials she carried over from Beirut.
2. See Ong, W.J. (1982). *Orality and literacy: The technologizing of the word.* London, UK: Routledge, pp. 71–74.
3. Following Gilles Deleuze and Pierre-Félix Guattari, Marks appropriates Alois Riegl's distinction between an optical and haptic image. Marks describes optical visuality as a way of seeing things from a distance to perceive them as distinct forms where the viewer is separated from the object (as "an all-perceiving subject"), whereas haptic visuality directly addresses the viewer's body in intimate ways: "It [haptic image] is more inclined to move than to focus, more inclined to graze than to gaze" (Marks, 2000, p. 162).
4. Stiegler shares a similar tone with Martin Heidegger's text *The Question of Technics* that set the technics as a way of "bringing forth" and criticizes its appropriation as a means-ends relationship in the modern era. Eventually, however, Stiegler puts more emphasis on the constitutive role of technics in its coupling with humans than Heidegger does.
5. In Heidegger, the past does not only refer to one's own past, but also the past that Being acquires through history and tradition without having lived it. This point explains both how Being is already temporally constituted and why Stiegler's tertiary memory is constitutive of primary and secondary retentions, and not vice versa.
6. As Westerkamp lays out, The Muzak Corporation got its start in the 1930s, supplying background music to industry (and settings like hotels, shopping malls, and offices in the later decades). Since the 1970s, Muzak has provided many other services including security and surveillance systems (1988, pp. 36–38).
7. Here, Westerkamp draws upon Jacques Attali's work *Noise: The Political Economy of Music* (1985) that tackles the political economy of the music industry.
8. See Foreman's article for a more in-depth discussion of Nancy's theoretical concepts.
9. See Nancy, J.-L. (2007). *Listening.* (C. Mandell, Trans.). New York, NY: Fordham University Press, p. 12.
10. See Dabashi, H. (Eds.) (2006). *Dreams of a Nation: On Palestinian Cinema*. New York, NY: Verso.
11. As Tawil-Souri notes, this is a deliberate reference to "present absentees", referring to the majority of Palestinians who were dispossessed from their land by Jewish forces during the Nakba. The transformation of Palestinian refugees into "present absentees" was institutionalized with Israel's 1948

Emergency Regulation Concerning Absentee Property, the 1950 Absentee Property Law, and the 1952 Citizenship Law.

12. Like Hatoum, Alsharif was born to Palestinian parents living in exile. She was born in Kuwait in 1985, but has had a nomadic life since then. Alsharif belongs to a younger generation of Palestinian woman artists of whom Hatoum was one of the pioneers. One of the early video works of Alsharif was titled *We Began by Measuring Distance* (2009), which can be read as a reference to Hatoum's earlier work *Measures of Distance* (1988).
13. Alsharif also produced a sound piece titled *Deep Sleep II* (2015), based on a brainwave generating binaural beats interwoven into field recordings from the Gaza Strip. From overcrowded streets to overgrown fields, drones in the sky, and birds flying by, it is meant to be an invitation to experience a place through its sounds. Within her nomadic position, Alsharif connects the viewer to larger histories of dispossession and inequality while interrogating what Gaza represents as a site, a conflict, and a metaphor. For further discussion on metaphors through which the realities of Gaza are obscured or exposed, see: Tawil-Souri, H. & Matar, D. (2016). *Gaza As Metaphor.* London: Hurst Publishers.
14. This quote is taken from an interview between Raquel Schefer and Basma Alsharif that took place in 2015. The text titled *The Impossibility of Describing a Place: An Interview with Basma Alsharif* was accessed on the website of the journal *La furia umana*, May 15, 2018: http://www.lafuriaumana.it/index.php/58-archive/lfu-25/399-raquel-schefer-the-impossibility-of-describing-a-place-interview-with-basma-alsharif

References

Alsharif, B. (2014). *Deep Sleep* [HD transfer from super 8mm film]. Video retrieved from https://vimeo.com/101257097

Foreman, I. (2011). Uncanny Soundscapes: Towards an Inoperative Acoustic Community. *Organised Sound, 16*(3), 264–271. https://doi.org/10.1017/S1355771811000276

Hatoum, M. (1988). *Measures of Distance.* [Video]. Western Front, Vancouver, BC, V5T 1S1, Canada.

Marks, L. U. (2000). *The Skin of the Film: Intercultural Cinema, Embodiment, and the Senses.* Durham, NC: Duke University Press.

Matar, D. (2011). *What It Means to Be Palestinian: Stories of Palestinian Peoplehood.* London/New York: I.B.Tauris & Co Ltd.

Naficy, H. (2001). *An Accented Cinema: Exilic and Diasporic Filmmaking.* Princeton, NJ: Princeton University Press.

Saloul, I. (2012). *Catastrophe and Exile in the Modern Palestinian Imagination: Telling Memories.* New York: Palgrave Macmillan.

Schafer, R. M. (1977). *The Tuning of the World*. New York: Knopf. Reprinted as *The Soundscape: Our Sonic Enviornment and the Tuning of the World*. Rochester, VT: Destiny Books, 1994.
Stiegler, B. (1998). *Technics and Time, 1: The Fault of Epimetheus* (R. Beardsworth & G. Collins, Trans.). Palo Alto, CA: Stanford University Press.
Stiegler, B. (2008). *Technics and Time, 2: Disorientation* (S. Barker, Trans.). Palo Alto, CA: Stanford University Press.
Stiegler, B. (2011). *Technics and Time, 3: Cinematic Time and the Question of Malaise* (S. Barker, Trans.). Palo Alto, CA: Stanford University Press.
Stiegler, B. (2014). *Symbolic Misery, 1: The Hyperindustrial Epoch* (B. Norman, Trans.). Palo Alto, CA: Stanford University Press.
Tawil-Souri, H. (2011). Qalandia Checkpoint as Space and Nonspace. *Space and Culture.*, *14*(1), 4–26. https://doi.org/10.1177/1206331210389260
Tawil-Souri, H. (2014). Cinema as the Space to Transgress Palestine's Territorial Trap. *Middle East Journal of Culture and Communication.*, *7*, 169–189. https://doi.org/10.1163/18739865-00702005
Truax, B. ([1984] 2001). *Acoustic Communication* (2nd ed.). Westport, CT: Greenwood Press.
Westerkamp, H. (1987). *Cool Drool* [Two-Channel Tape]. Audio Recording. Retrieved from https://www.sfu.ca/~westerka/program_notes/cooldrool.html
Westerkamp, H. (1988). *Listening and Soundmaking: A Study of Music as Environment*. Unpublished Master's Thesis, Simon Fraser University, Vancouver, BC, Canada.

CHAPTER 11

Responsive Listening: Negotiating Cities of Sirens, Smartphones and Sensors

Sarah Barns

Introduction

Urban life today is shaped, more and more, by our interactions with digital devices, infrastructures and platforms. Cities are now being designed in a way that is 'smart', utilising distributed sensors, algorithmic governance and machine-learning techniques to support closer monitoring and management of urban infrastructures, atmospheres and services—from traffic monitoring and charging to energy and water utilities, to air quality and crowd management. The computational intensification of urban life has, in turn, unleashed a wave of new thinking about the potentials of digitally responsive cities, and the appropriate design and governance frameworks for cities that are being 'run on information' (Goldsmith & Crawford, 2014). This pervasive integration of digital technologies within the physical fabric of cities is also taking place at a time when smartphones have become central devices for navigating, documenting, connecting and sharing experiences of urban life. If the lines between digital and physical spaces are becoming porous, as smartphones and sensors proliferate, so the informational ecologies of digital cultures, centred around pervasive social media platforms, are becoming more and more a part of our every-

S. Barns (✉)
University of New South Wales, Sydney, NSW, Australia
e-mail: sarah@sitelines.media

M. Droumeva, R. Jordan (eds.), *Sound, Media, Ecology*, Palgrave Studies in Audio-Visual Culture,
https://doi.org/10.1007/978-3-030-16569-7_11

day interactions with the world around us—what I have elsewhere described as a kind of 'platform urbanism' (Barns, 2016).

In the light of these developments, this chapter explores how practices of contemporary listening negotiate different kinds of urban informational and infrastructural technologies. With interest in listening practices growing as a vital, if often neglected, aspect of urban life (Beatley, 2013), I draw here from the work of sound historians who explore historical changes in listening practices through time, absorbing the urban disruptions wrought by industrialisation and mass remediation. In turn, I situate today's digitally responsive cities within a longer historical trajectory of urban technological disruption and remediation, in order to explore the different ways in which technologies of listening are today seeking to modulate and to 'make sense' of urban environments. Where the modern soundscape, in Schafer's original conception, was associated with the rise of noise pollution and reverberatory technologies, new technologies of sensing and modulation invite us to reflect on a diversity of different listening practices that respond to an increasingly informationalised soundscape.

Looking to the future, with urban spaces endowed with more machinic listening capabilities, I explore how computational devices are now being tasked with 'active listening' capabilities. Soundscape analyses of cities, which we might think of as computational 'earwitness' accounts, are being used to generate proxy or human-like responses to sonic events through the use of big data or through the use of tiny microphones used by smart phones and other smart devices. If, for decades, active listeners and sound practitioners have been inspired by the use of mobile media to recalibrate the listening experience as one of multimodality, historical contingency, ecological awareness or aesthetic desire, these technology developments point to the proliferation of non-human actors entering the field over the coming decades. In this context, new modalities of acoustic ecology emerge, in which artificial modes of (sonic) intelligence contribute to analyses of contemporary soundscapes, augmenting not only what we hear but also how we listen, in both embodied and computational ways.

Learning to Listen to the City

The city is starting to be heard. Of course, cities are inherently noisy places, brimming with the sounds of sirens, traffic, whistling winds, jack hammers, chatter, bird song and pouring rain. And yet, this is not the way cities have been imagined and represented across architecture and urban design fields.

As Blesser and Slater (2007) ironically asked, "Spaces speak. Are you listening?" As it is, the work of designing cities, precincts and buildings has tended not, in fact, to pay much attention to the way a space might be designed sonically, and tends often to obscure the quality of sound as a factor in the way we experience place and space. Indeed, as has been widely recognised, an emphasis on visual methods through which to plan, design and interpret city spaces has historically relied on the cultural superiority of seeing as a privileged mode of urban experience (Leach, 2002; Pallasmaa, 1996; Rodaway, 1994; Zardini, 2005). As Mattern and others have noted, scholarly attention towards the nature of urban, particularly where it focuses on forms of technological mediation in the city, has likewise typically situated the urban subject within theories of the gaze and the spectacle, addressing literary, cartographic, photographic and filmic representations of the textual, visual city (Boyer, 1996; Bruno, 2002; Mattern, 2008, p. 158; McQuire, 2008).

Some argue that this has not simply been a matter of neglect: Zardini (2005, pp. 20–21) has argued that by privileging visual perception in capturing the qualities of urban space, the techniques of architecture and urban planning treat sounds (and odours) as highly disturbing elements. Others see the visualism inherent in urban design practices as reflecting a deeper instrumentalism prevalent across Westernised ways of knowing and perceiving the world—what Heidegger (1977) calls the *Gestell*: a systematic visual enframing of the world as a separated object of knowledge. In response to these ideas, a more 'sensory urbanism' has emerged, seeking to foreground a multisensory experience of place, amplifying more embodied, and situated, accounts of cities, alive to the affective qualities of place and space (Rodaway, 1994; Rykwert, 2000). This sensory turn has been a part of the wider resurgent interest in the practice of listening as a vital, if neglected, dimension of subjective urban experience (Smith, 2004). Drawing on Schafer's concept of the soundscape, which reimagined the world as a 'macro-cosmical' or vast musical composition (Schafer, 1977), sound historian Emily Thompson sought to recover the construction of a modern soundscape, consisting "not only of the sounds themselves, the waves of acoustical energy permeating the atmosphere in which people live, but also the material objects that create, and sometimes destroy, those sounds" (Thompson, 2004, p. 1). This resurgent interest in the listening self has amplified the many historical contingencies shaping our modern auditory. Conner (1997) used the notion of the modern auditory to describe the co-constitution of technologies of amplification, and material environments that comprises the soundscape, as experienced through the practice of listening. For Conner, the most important, distinguishing feature of the modern auditory as compared to earlier auditory conditions, is its capacity to disintegrate and reconfigure space (1997, p. 206). Many others, including Sterne (2003), Chion (1994) and Altman (1992), have likewise traced different cultures of listening as they have been shaped by diverse technologies, whether technologies of urban transformation and industrialisation or electronic amplification and reproduction.

Through these accounts we gain unique insights into the historical construction of our soundscape as an intimate interplay between changing technologies of the city and the way we listen and experience the world sonically. In studying the rise of electronic amplification, Sterne came to resist the idea that hearing was a kind of manifestation of 'pure interiority' (2003: 15) or what Eisler and Adorno felt the 'pre-individualistic' traits of listening were, compared to the instrumentalist nature of vision (in Sterne, 2003, p. 74). He describes the act of listening as "a directed, learned activity: it is a definite cultural practice", rather than what McLuhan once suggested might be expressive of "nature in the raw" (in 2004, p. 71).

Returning to the sounds of cities of the early twentieth century also led historian Karin Bijsterveld (2001) to conclude that the sound of technology, and the way we experience it, is a key dynamic of contemporary technological culture. In her study of noise abatement campaigns between 1900 and 1940, Bijsterveld examined how rapid industrialisation drastically changed the sonic environment of modern cities. At the time, these changes were experienced physically as a profound dislocation of the senses. Thompson quotes a *New York Times* reporter from 1926 who put it simply:

> The machine age has brought so many new noises into existence, the ear has not learned how to handle them. It is still bewildered by them. (2002, p. 195)

The "diabolical symphony of the mechanical age", as Bijsterveld (2001) describes it, was experienced as a peacetime version of shell shock. Every new invention created new sounds, which in turn became the topic of heated controversy. Thompson's study of noise abatement campaigns in New York also evidence how little the sound world circa 1930 had in common with that of 1900 (Thompson, 2004, pp. 190–91). Protests against the interventions of industrial-era noise saw the emergent modern soundscape as a costly threat to the health and efficiency of urban citizens, which would surely "shorten the life of countless sufferers" (Thompson, 2004, p. 174).

The rise of architectural acoustics in the 1930s helped isolate citizens sonically from the world outside. Thompson (2002) traced how innovations in building design, in particular the use of sound-absorbing materials, created the conditions for new interior, modern soundscapes which essentially banished the sounds of the (increasingly noisy) urban soundscape. In this way, listening practices were reconfigured to respond differently to different spaces: interior spaces were sound absorbing, while exterior spaces were left exposed as a 'diabolical symphony'.

Technologies of amplification also dramatically changed the way we listen in public spaces. Loudspeakers, headsets, headphones, telephones and phonographs meant mediated sounds became more and more central to the ecology of urban life (Bull, 2004, p. 104, pp. 237–238; Sterne, 2003). Adorno describes the penetration of mediated sounds in the city as a normalisation of the soundscape, in which "everything is to sound like the radio" (Adorno, 1991, p. 58). Noticing the way radio had transformed the domestic spaces of Berlin in the 1920s, Kracauer saw people left in a state of "permanent receptivity", exclaiming "Who could resist the temptation of those dainty headphones? [...] Silent and lifeless, people sit side by side as if their souls were wandering far away"—more attuned to the playground of worldwide (broadcast) noises than their local environs (Kracauer, 1995, p. 333). For the radio pioneer Marinetti, radio enabled "the immensification of space: no longer visible and frameable, the stage becomes universal and cosmic" (as cited in Conner, 1997, p. 208).

Innovations in mediated sound—whether cosmic or domestic—therefore facilitated a kind of displacement, encouraging new ways of listening that distracted people from more intimate and immediate forms of social engagement. Schafer characterised this condition as "schizophonic", resulting from sounds being "torn from their natural sockets and given an amplified and independent existence" (1973, p. 90). As a consequence, "any sonic environment can now become any other sonic environment": we are less likely to experience the unique 'soundmarks' of a space (Schafer, 1977, p. 100). Sounds can be dislocated not only in space but also in time: "A record collection that contains items from widely diverse cultures and historical periods would seem, to a person from any century but our own, an unreal and surrealistic juxtaposition" (p. 34). Electro-acoustics are in this sense also understood as promoting an 'acousmatic' experience of sound: an experience of listening in which the source of sound is without visual anchor, 'hidden' from the listener, whether in time or space (Blesser & Slater, 2007, p. 182; Conner, 1997).

And yet, as scholars and historians of sound emphasise, listening is rarely an autonomous experience. It acts in a way that Douglas Khan describes as "always relational, a constant deflection" (Khan, 1994, p. 15). This mutability and relationality also means that sound can act as a privileged figure of sensory interchange; the 'listening eye' for example, facilitates a gaze mutated to the conditions of hearing. For the architect Pallasmaa (1996), the mutability of sound acts as a constructive resource. Just as the qualities of concrete and steel might be manipulated by designers to promote a satisfactory architectural composition in space, Pallasmaa has advocated the deliberate manipulation of sound quality to promote an intimate, multisensory experience of place.

Augmented Listening Technologies

In more recent years, the urban soundscape has been increasingly modulated by distributed technologies of listening—from the personal stereo to the ubiquitous iPhone, with its white earbud headphones. These have, once again, brought about fundamental shifts in the subjective orientations of urban citizens in public space (Barns, 2014; Beer, 2007; Biggs, 2008; Bull, 2007; Droumeva, 2016; Thibaud, 2003; Tonkiss, 2003). The ethnographic work of Michael Bull has explored mobile listening devices as a new kind of 'auditory looking' that is tantamount to a "radical auditory re-prioritisation of forms of urban experience" (Bull, 2007, p. 5). To Thibaud, 'walkman listeners' create new sonic territories, recomposing the city through spatio-phonic behaviours (2003, p. 329).

Studying urban cultures of iPod use in the early 2000s, Bull aimed to 'make visible' the centrality of personal stereos to urban experience today. Interviewing mobile listeners, Bull found that by creating more privatised auditory experiences, they were able to get 'more' out of their environments, not by interacting socially, but precisely by not interacting. "Not bored" a listener called 'Donna' tells Bull, "it livens everything up. Everything's on a higher level all the time. It makes it seem a bit busier. You get excited. Everything's happening" (2007, p. 36). Through the 'pleasurable sound bubbles' created by a personal listening soundtrack, Bull sees listeners curating different kinds of urban experience, but in doing so they shut off the 'real' sounds of the city, which are, in turn, relegated to the disenchanted and mundane outside world (Bull 2007: 36). In Thibaud's account, turning on a private soundscape begins when one leaves the house. It is paradoxical, he writes, that "access to public space is associated with the beginning of private listening, whereas entering the private realm is associated with [...] openness towards the surroundings" (2003, p. 331).

Mobile practices of mediated listening have provoked both critical and creative responses. To urban sociologists, these practices constitute a kind of withdrawal from the world, or what Fran Tonkiss has called a "social deafness" (2003, p. 304). Bull described this form of listening as a highly aestheticised way of experiencing the city, in which the listener is empowered to harmonise or 'curate' an experience to suit one's mood and one's personal desire. To David Beer, a sociologist of urban technology, this kind of mediated listening is not so much about the distancing of listeners from their environments, but is simply part of an everyday negotiation between different types of technologically constituted soundscapes. He writes that "rather than being lifted *out of* aural ecologies, mobile listeners are constantly and inescapably integrated or networked in these territories" (2007, p. 859; emphasis added). In other words, listening remains a process of negotiating the multilayered, networked informational structures of the city, through technologies of mediation that are more and more co-constitutive of the very fabric of everyday existence (Droumeva, 2016). Sound composers and practitioners have embraced the mobile medium to creatively engage multiple layers of meaning and relationality between self and space. As Barclay (2016) has written, the proliferation of mobile devices has created many new opportunities through which to explore our social, cultural and ecological environments through sound. Rather than reducing the potential for more 'authentic' engagement with urban spaces though embodied listening, in Barclay's account mobile listening is actively integrated into its everyday informational infrastructures. If sound composition using locative media has now been a vibrant field of soundscape design for some decades (Barclay, 2016; Butler, 2006; Droumeva, 2016; Hight, 2004; Pinder, 2005), the more recent proliferation of ubiquitous, everyday mobile platforms has in turn opened up more diverse audiences and creative potentials.

To Barclay (2016, p. 21) working with locative media through sound has the potential to reconnect people with diverse ecosystems, including natural ecosystems, rather than simply alienating them. Sound practitioners who work with augmented, mobile and site-specific platforms use sound composition as a way to open up multiple experiential and affective interactions with a site. In doing so, they are curating and recombining diverse sonic informational resources to create a sound-led synesthetic experience of a space's plurality and permeability. Oral histories, field recordings, narrative voice and music compositions combine, through site-specific soundscape design, to facilitate immersive experiences of a site in ways that recalibrate the very nature of socio-spatial encounter.

Such forms of immersion are ultimately, and powerfully, driven by sound. Those such as Thibaud (2003) and Biggs (2008) see this work not only as soundscape design but as a form of multimodal engagement, in which curated soundscapes are used to compose a kind of 'physical cinema', creating a 'city screen' where the material world forms a quasi-cinematic image accompanying a mobile soundtrack. For sound practitioners, the practice of listening in this way constitutes a way to actively engage the informational resources of a city's past and present, its intimate and public, machinic and biological patterns. Listening in this way, the city becomes a vast recorded archive, to be recomposed in ways that amplify a site's historical contingency (Barns, 2011). The ubiquitous nature of mobile use in cities today also means the way we listen is made up of a set of deliberate choices about one's orientation to the world. Listening to news, podcasts, music and social media videos are all a part of the everyday auditory space of urban citizens. Ubiquitous devices not only enable listeners to curate musically their everyday interactions with the city, but can also be used to map and document locational sounds.

While mobile audio devices have been with us for some decades, the introduction of in-ear computing technologies into personal headphones allows the urban soundscape to be modulated in more subtle ways. Doppler Labs' 'Hear One' headphones, for example, offered listeners a way to 'change the volume of the real world', an 'always-in' way to equalise and audio-process everyday sounds, allowing listeners to use audio filters to determine their own acceptable frequency limits (Pierce, 2017). Another headphone, the 'nuraphone', generates a sonic profile of a listener's ears, which enables the music they play to match their "hearing system" (Nuraphone, n.d.). IQbuds, wireless smart hearing buds by Nuheara, are another headphone that seek to "overcome the lack of situational awareness" that comes with normal headphones by "integrating digital and physical worlds", and selectively tuning out specific conditions of a listener's audio environment (Nuheara, n.d.).

While sensor-driven, in-ear audio-processing capabilities have been with us for some time, in the form of hearing aids, this new generation of 'smart hearing' devices is now being promoted as 'the next great frontier' of augmented reality. In-ear computational sensing targets the psychological effects of auditory disturbances, the physical effects of hearing loss and the desire on the part of consumers to exercise greater control over the way they interact with others and their environment (Nuheara, n.d.). And yet, real-time, in-ear computational hearing has also been described as a sort of sensory dislocation. As one review of the ill-fated Hear One headphones reflected: “Audio processing in my ears all day long gave me some sort of weird uncanny valley feeling. I felt present but absent. It was a bit like wearing Google Glass, or a VR headset all day” (Stein, 2017).

The ‘uncanny valley’ feeling this reviewer reported perhaps echoes the sense of dislocation experienced in the early days of telephone usage in the late nineteenth century, when the device was popularised as a means to listen to live musical performances known as ‘telephone occasions’ (Marvin, 1988, p. 222). Or perhaps, if indeed they do grow in popularity, such in-ear devices may provoke more and more creative responses, in which soundscape design integrates ambient and musical properties of the environment and personal soundtrack in more and more integrated ways. As we have heard, listening practices are not static, but evolve and adapt in response to different technologies and cultures of auditory reproduction and remediation.

Listening to the City’s Data Exhaust

Each day our cities generate huge volumes of data, sometimes referred to as a new kind of ‘data exhaust’, generated through our everyday mediated interactions, transactions and navigation. While digital participation tends to be framed in terms of active interactions—speaking, ‘produsing’ (Bruns, 2008) and creating evermore big data—Kate Crawford reminds us that digital participation can also constitute a kind of listening. Crawford (2009) describes states of moving between ‘listening in’ and commenting online, whereby listening constitutes a vital part of the activity of online communities, particularly through social media.

Diverse ‘drifting and differentiating modes’ of digital communications, as Terranova describes them (2004, p. 52), enacted through our engagements with digital platforms, are now amplified through a new infrastructure of listening in the city—a city of sensing technologies. Proliferating our everyday spaces are many billions of connected things, known as the ‘internet of things’ (IoT). Footpaths, parking meters, surveillance cameras, light poles, traffic lights, bridges, trees, water, air quality, temperature: each is now subject to data-driven methods of monitoring, sensing and managing. In turn, computational methods of analysis are now being deployed to understand the contemporary urban soundscape. The vast networked infrastructures of our digitally infused urban fabric are beginning to listen to the city, too.

Researchers at the Centre for Urban Science and Progress (CUSP) at NYU have been investigating the use of digital micro-electromechanical systems microphones (MEMS) to monitor noise pollution in New York (Mydlarz, Nacach, Park, Roginska, 2014). There are four MEMS in every Apple iPhone, but they are also increasingly used in IoT applications, from smart phones to smart home devices, to security monitors, to 'fitbits' and smart parking meters. CUSP researchers approach New York as a kind of 'living lab', a city to both test the potentials of digital MEMs for urban monitoring, but also to address the worldwide problem of noise pollution (2014, p. 2). If noise pollution is the ultimate problem CUSP researchers are trying to solve, their primary concern is with using sensor networks to understand spatio-temporal acoustics itself, in ways that support measuring, streaming, archiving, analysing and visualising urban soundscapes (Park et al., 2014).

A similar project called *Sensor City*, based in Assen in the Netherlands, deployed hundreds of static sensor units with acoustic monitoring equipment. This project aimed to introduce more 'human-like' understandings of the urban soundscapes as captured by sensors; in other words, to correlate the data generated by these sensors to human responses to particular 'sound events' and sonic profiles. For another project, *Making Sense of Sounds* (n.d.), the following question remains: can machine systems be taught to categorise and respond to sound events like a human? The billions of tiny sensors distributed throughout our cities today can, in their own way, be understood as listening to the contemporary soundscape. Machinic listening—better known as machine learning—uses the data generated by MEMs sensors to structure and analyse data for artificial intelligence (AI) applications. At present, transmission, storage and processing costs of MEMs data are limiting the AI applications of this acoustic data. Yet the future capacity to pair AI and MEMS sensors into a symbiotic system is considered the gateway to a "new world of undreamt-of opportunities for [AI] designers and end users" (Gandhi, 2018).

In this not-so-fanciful future world, computational forms of listening can be anticipated to symbiotically respond to the soundscapes of cities, generating, via artificial intelligence applications, new digital products and services that seek to integrate and envelope us evermore intimately into their informational ecologies. We are listening—to our cities, to our digital networks—and so are our devices and our infrastructures. No longer inert, the physical fabric of the city is, in the words of smart city researcher Anthony Townsend, "waking up and becoming aware of itself" (2009, p. 18). We are each, in our different ways, responding, recombining and restructuring data feeds according to our own native languages and signal processes.

Conclusion

In this chapter I've explored listening as a site of negotiation between different kinds of urban informational and infrastructural technologies. The emergence of a more sensory urbanism, combined with an appreciation of the historical contingencies of today's modern soundscape, has led to deeper engagement with historical waves of acoustic disruption in the city, including disruptions wrought by industrialisation and the rise of new recording and reproduction technologies. Situating the emergence of digitally responsive cities within this historical trajectory of technological disruption and remediation, I've explored the way different technologies of listening in the city today can be seen to facilitate diverse practices of active listening, particularly as acoustic sensing technologies are used to modulate and make sense of urban environments—in the form of machinic 'earwitness' accounts.

This modern auditory, which includes digitally enraptured selves and embedded sensors, provokes the spectre of an integrated informational ecology in the not-too-distant future, in which subjective experience is more tightly conditioned and regulated through data-driven modes of sensing and communicating. In this context, computational modes of listening may be a more central part of hearing cultures. Instead of being reduced to a more functional landscape of big data and digital-informational infrastructures, the acoustic-informational ecology of the smart city might be explored as a complex field of cognition, computation, desire and experience (Gandy, 2005, p. 41). Such an acoustic ecology encompasses not only our subjective selves as earwitnesses but also sites of data accumulation, acoustic design and curation, and biophysical feedback—in other words, a vast array of vibrating, listening, sound-emitting, *macro-cosmical* subject-objects.

And so, as Schafer reminds us, "the soundscape of the world is changing" (Schafer, 1977), but so too is the relationship between listeners and their environments. Cultures of mediated listening, encouraged by the growing sophistication of augmented hearing and the proliferation of smartphone use, means many are habituated to a highly 'schizophonic' experience of their environments. In this context, listening practices encompass diverse informational, industrial and ecological dimensions of a soundscape. And yet the question remains: how does acoustic ecology as a field respond to the computational intensification of everyday lifeworlds? Modulated by technologies of augmentation and machine learning, the contemporary soundscape is subject to more and more heterogeneous ways of listening. If acoustic ecology emerged as a restorative practice, one that drew from hearing as a more 'interior' way of experiencing the world and as a way to amplify the polluted nature of our soundscapes, the coming years will surely see acoustic ecology also evolve as a practice that advocates for the closer appreciation of the embodied, situated self. In so doing, we may see the field raging against not only the pollution of the natural world, but also emerging forms of 'digital enclosure' (Andrejevic, 2009), and the monetisation of human attention enabled by evermore sophisticated technologies of personalisation and quantification.

References

Adorno, T. (1991). *The Culture Industry: Selected Essays on Mass Culture*. London: Routledge.

Altman, R. (1992). Introduction: Sound/History. In R. Altman (Ed.), *Sound Theory/Sound Practice*. New York/London: Routledge.

Andrejevic, M. (2009). Privacy, Digital Exploitation, and the Digital Enclosure. *Amsterdam Law Forum, 1*(4). Available at http://amsterdamlawforum.org/article/view/94/168. Accessed Oct 2018.

Barclay, L. (2016). Augmenting Urban Space with Environmental Soundscapes and Mobile Technologies. *Soundscape: The Journal of Acoustic Ecology, 16*, 21–34.

Barns, S. (2011). Street Haunting: Sounding the Invisible City. In M. S. Foth (Ed.), *From Social Butterfly to Engaged Citizen* (pp. 203–217). Cambridge, MA: MIT.

Barns, S. (2014). Sounds Different: Listening to the Proliferating Spaces of Technological Modernity in the City. *Space and Culture, 17*, 4–15.

Barns, S. (2016). Visions of Urban Informatics: From Proximate Futures to Data-Driven Urbanism. *Fibreculture Journal, 29*.

Beatley, T. (2013). Celebrating the Natural Soundscapes of Cities. *The Nature of Cities*. Available at https://www.thenatureofcities.com/2013/01/13/celebrating-the-natural-soundscapes-of-cities/. Accessed Oct 2018.

Beer, D. (2007). Tune Out: Music, Soundscapes and the Urban Mise-En-Scene. *Information, Communication & Society, 10*, 846–866.

Biggs, B. (2008). *Like It Was a Movie: Cinematic Listening as Public Art* (Working Paper 36). Princeton, NJ: Princeton University Centre for Arts and Cultural Policy Studies.

Bijsterveld, K. (2001). The Diabolical Symphony of the Mechanical Age: Technology and Symbolism of Sound in European and North American Noise Abatement Campaigns, 1900–40. *Social Studies of Science, 31*, 37–70.

Blesser, B., & Slater, L. (2007). *Spaces Speak – Are You Listening? Experiencing Aural Architecture*. Cambridge, MA: MIT Press.

Boyer, M. C. (1996). *Cyber Cities: Visual Perception in the Age of Electronic Communication*. New York: Princeton Architectural Press.

Bruno, G. (2002). *The Atlas of Emotion: Journeys in Art, Architecture, and Film*. New York: Verso.

Bruns, A. (2008). *Blogs, Wikipedia, Second Life and Beyond: From Production to Produsage* (Digital Formations 45). New York: Peter Lang.

Bull, M. (2004). Sound Connections: An Aural Epistemology of Proximity and Distance in Urban Culture. *Environment and Planning D: Society and Space, 22*, 306–116.

Bull, M. (2007). *Sound Moves: iPod Culture and Urban Experience*. London/New York: Routledge.

Butler, T. (2006). Memory-Scape: Using New Technology to Integrate Art, Landscape and Oral History. *Rising East Online* No. 5.

Chion, M. (1994). *Audio-Vision: Sound on Screen*. New York: Columbia University Press.

Conner, S. (1997). The Modern Auditory. In R. Porter (Ed.), *Rewriting the Self: Histories from the Renaissance to the Present* (pp. 203–223). London/New York: Routledge.

Crawford, K. (2009). Following You: Disciplines of Listening in Social Media. *Continuum: Journal of Media and Cultural Studies, 23*(4).

Droumeva, M. (2016). Curating Aural Experience: A Sonic Ethnography of Everyday Media Practices. *Interference: A Journal of Audio Cultures, 5*, 72–88.

Gandhi, K. (2018). *AI and Mems Sensors: A Critical Pairing*. SEMI. Available at http://blog.semi.org/technology-trends/ai-and-mems-sensors. Accessed 1 Apr 2019.

Gandy, M. (2005). Cyborg Urbanization: Complexity and Monstrosity in the Contemporary City. *International Journal of Urban and Regional Research, 29*(1), 26–49.

Goldsmith, S., & Crawford, S. (2014). *The Responsive City: Engaging Communities Through Data-Smart Governance*. New York: Wiley.

Heidegger, M. (1977). The Age of the World Picture. In M. Heidegger (Ed.), *The Question Concerning Technology and Other Essays* (pp. 115–154). New York: Harper and Rowe.

Hight, J. (2004). *Narrative Archaeology: Reading the Landscape*. Available at http://web.mit.edu/comm-forum/legacy/mit4/papers/hight.pdf

Khan, D. (1994). Introduction: Histories of Sound Once Removed. In G. Khan DaW (Ed.), *Wireless Imagination: Sound, Radio and the Avant-Garde*. Cambridge, MA: MIT Press.

Kracauer, S. (1995). *The Mass Ornament: Weimar Essays*. Cambridge, MA: Harvard University Press.

Leach, N. (2002). *The Hieroglyphics of Space: Reading and Experiencing the Modern Metropolis*. London: Routledge.

Marvin, C. (1988). *When Old Technologies Were New: Thinking about Electric Communication in the Late Nineteenth Century*. New York/Oxford: Oxford University Press.

Mattern, S. (2008). Silent, Invisible City: Mediating Urban Experience for the Other Senses. In F. Eckardt, J. Geelhaar, L. Colini, K. Willis, K. Chorianopolous, & R. Hennig (Eds.), *Media City: Situations, Practices, Encounters* (pp. 155–176). Berlin, Germany: Frank & Timme.

McLuhan, M. (2004). 'Visual and Acoustic Space.' In Cox, C. and Warner, D. (eds) *Audio Culture: Readings in Modern Music*, pp. 67–72. London: Continuum Publishing Group.

McQuire, S. (2008). *The Media City: Media Architecture and Urban Space*. London: Sage.

Mydlarz, C., Nacach, S., Park, T. H., & Roginska, A. (2014). *The Design of Urban Sound Monitoring Devices*. Los Angeles, CA: Audio Engineering Society (AES) Convention 137.

Nuheara. (n.d.). *Nuheara Homepage*. Available at https://www.nuheara.com/. Accessed 1 Apr 2019.

Nuraphone. (n.d.). *How It Works*. Available at https://www.nuraphone.com/pages/how-it-works

Pallasmaa, J. (1996). *The Eyes of the Skin: Architecture and the Senses*. London: Academy Editions.

Park, T. H., Turner, J., Musick, M., Lee, J. H., Jacoby, C., Mydlarz, C., et al. (2014). *Sensing Urban Soundscapes* (EDBT/ICDT Workshops) (pp. 375–382).

Pierce, D. (2017). Inside the Downfall of Doppler Labs. *Wired Magazine*.

Pinder, D. (2005). Arts of Urban Exploration. *Cultural Geographies, 12*(4), 383–411.

Rodaway, P. (1994). *Sensuous Geographies: Body, Sense and Place*. London: Routledge.

Rykwert, J. (2000). *The Seduction of Place: The History and the Future of the City*. Oxford: Oxford University Press.

Schafer, R. M. (1973). *The Music of the Environment* (No. 1 of an Occasional Journal Devoted to Soundscape Studies). Vienna, Austria: Universal Edition.

Schafer, R. M. (1977). *The Tuning of the World*. New York, NY: Knopf. Reprinted as *The Soundscape: Our Sonic Environment and the Tuning of the World*. Rochester, VT: Destiny Books, 1994.

Smith, M. (2004). *Hearing History: A Reader*. Athens, Greece: University of Georgia Press.

Sounds MSo. (n.d.). *Making Sense of Sounds*. Available at http://cvssp.org/projects/making_sense_of_sounds/site/about/

Stein, S. (2017). Here One Buds Could be the Future of Augmented Listening. *CNet*.

Sterne, J. (2003). *The Audible Past: The Cultural Origins of Sound Reproduction*. Durham, NC/London: Duke University Press.

Terranova, T. (2004). *Network Culture: Politics for the Information Age*. London: Pluto Press.

Thibaud, J. (2003). The Sonic Composition of the City. In M. Bull & L. Back (Eds.), *The Auditory Cultures Reader* (pp. 329–341). Amsterdam: Berg.

Thompson, E. (2002). *The Soundscape of Modernity Architectural Acoustics and the Culture of Listening in America 1900–1933*. Cambridge, MA: MIT Press.

Thompson, E. (2004). Shaping the Sound of Modernity. In M. Smith (Ed.), *Hearing History: A Reader* (pp. 341–348). Athens, Greece: University of Georgia Press.

Tonkiss, F. (2003). Aural Postcards: Sound, Memory and the City. In M. Bull & L. Back (Eds.), *The Auditory Culture Reader* (pp. 303–309). New York: Berg.

Townsend, A. (2009). Forward. In M. Foth (Ed.), *Handbook of Research on Urban Informatics: the Practice and Promise of the Real-Time City*. Hershey, PA: Information Science Reference, IGI Global.

Zardini, M. (2005). Towards a Sensorial Urbanism. In M. Zardini (Ed.), *Sense of the City: An Alternate Approach to Urbanism* (pp. 17–27). Montreal, Canada: Canadian Centre for Architecture/Lars Muller Publishers.

CHAPTER 12

Listening to Traffic with Guts and Antennae

Andra McCartney

When I talk with other listeners, there is always conversation about road traffic. Other topics come and go, but this one seems ubiquitous, like the sound in our daily lives. As Karin Bijsterveld (2008) points out, traffic noise has been a common complaint throughout the history of noise regulation. From acoustic ecology to film production, field recordists work to minimize the constant drone of traffic, using bass roll-off filters. Our minds work to ignore the sound or transform it into something less repetitive and predictable.

The attention and conversation encouraged in soundwalks and listening workshops result in heightened listening to and reflection on the vibrations and flows of road traffic, the most constant sounds of many places. Roads and their margins are ecotones or landscapes of tension where intersecting habitats of plants, animals, cars, bicycles, motorbikes, trucks, trailers, recreational vehicles, pedestrians, beasts of burden and streetcars are ordered by laneways, held and released by traffic signals, covered with dust, gravel, tarmac, and suffused with the whines, roars, thrums, and tickings of passing vehicles, as well as their spectacle, smells, and toxic emissions. How do resonant bodies sense vibrations of traffic and make sense of them? What kinds of prostheses and practices are used

A. McCartney (✉)
Concordia University, Montréal, QC, Canada
e-mail: andra.mccartney@concordia.ca

M. Droumeva, R. Jordan (eds.), *Sound, Media, Ecology*,
Palgrave Studies in Audio-Visual Culture,
https://doi.org/10.1007/978-3-030-16569-7_12

to do this, and how do these technological processes affect daily sonic experiences of traffic, landscape design that pays attention to sound, and the creation and reception of sonic artworks involving traffic?

These questions arise out of sound research projects conducted at the Communication Studies department in Montreal, Quebec, at Concordia University, from 1999 to 2015—*Soundwalk to Home* (1999–2002), *Journées Sonores canal de Lachine* (2000–2005), and *Soundwalking Interactions* (2006–2015)—and are part of an extended series of creative works on traffic. I discuss research-creation practices from these projects such as observations in sound measurement using decibel meters, sonic experience during soundwalks and other listening activities, and exploratory conversations with participants and online audiences, which became inspiration for further thought and speculation about traffic issues. Listening tactics for living with and transforming traffic sound are suggested by participants in these research projects, some of which resonate with similar tactics used by artists working with traffic sound. The research questions were further developed through an invitation from Zeynep Bulut to include a sound piece and talk in the symposium Resonant Bodies: Landscapes of Tension (ICI Berlin, 2013). In that context, researchers were encouraged to think about points and flows of tension in different kinds of landscapes, including within the terms that we use, such as silence, sound, soundscape, and noise.

What does the sound of traffic mean, and what points and flows of tension does it express? In Hildegard Westerkamp's *Kits Beach Soundwalk*, she speaks of the monstrous sound of the city. When I listen to the piece, what I hear sounds like a drone of distant traffic; less the sounds of buildings in the city, more the sonic emissions of vehicles and routes that connect a suburban beach to the center. This is an important sound in the piece that forms a fulcrum between interfering city and intimate nature with its doorway to imagination. The sound is pushed back to reveal nature through filtering, just as field recordists regularly use bass roll-off filters to diminish traffic sound.

But the sound of traffic does not only mean city. It is not only urban. It is also the sound of the rural highway village where I live, and I heard it in a wilderness park as audience members drove through the forest to an environmental performance. It is the sound that acoustic ecologists make on their way to study and record in pristine far-away locations. The sound of traffic means how, and how much we value mobility. Traffic noise despite its ubiquity is not well understood in discourse about noise mitigation, because much of its power and volume lies in the infrasonic or low-frequency range that does not register on the dBA meter that is always and unquestioningly used for noise measurement regulation (Garon-Sayegh, 2018; St. Pierre & Maguire, 2004). While doing research for a sound project around the Lachine Canal in Montreal, we took decibel measurements while repeatedly visiting locations near the canal, using both dBA and dBC weighting. The dBA weighting mimics how human ears hear, while the dBC weighting also measures the low frequencies, the rumblings that our guts perceive. The largest differences that we found between the two measurement scales were in the area where the canal park follows a multi-lane highway into Montreal. Including low frequencies gave a level 20 dB higher, hidden highway sound.

Ideas about noise measurement have been persistently linked to the ear and to sounds that can be pinpointed for legal responsibility: low-frequency sounds are hard to localize and so it is more difficult to assign individual blame. Scant attention has been paid in noise measurement until recently to how bodies listen with their guts, and what are the effects on bodies of sounds that are more felt than heard. But the type of sound also matters in this discourse, beyond frequency and amplitude. My recent online searches on infrasonic sound level measurement resulted in far more sites about windmills than about road traffic. This is more than a politics of frequency or of loudness. Some listening activities make us more aware of our position *as* traffic. During soundwalks for instance, traffic flows become points of articulation, discussion, and reflection. During planning, questions of access to a public soundwalk involve traffic issues. As Andrew Wilson points out, wilderness parks are often difficult to travel to by public transit, and routes to some were even designed with bridges that would exclude public buses (1998). Hiking trails are hard to access by stroller or wheelchair. To create a walk that is accessible to these vehicles, a soundwalk designer must often decide against quiet wilderness locations.

Soundwalkers *become* traffic, can stop traffic with their presence as a group, walking and listening intently, restless, shape-shifting. Our flocking bodies shape sensations, texture and pacing of movements through space, inflecting each other's sonic experience, hearing punctuations of footwear, clothes, and phones, matching the pace of the slowest, following the leader, fanning out, splitting up, flowing together again. Soundwalkers move through and around different kinds of roads: the chaotic, noisy, and diverse roadways of New Delhi, the highly regulated and diverse lanes of Amsterdam, the individualized, speedy, and homogenized traffic of major highways, the highs and lows of tourist traffic toward cottages and parks, the far-off whisper of traffic heard from rural retreats. Listeners in Kinshasa, Democratic Republic of the Congo, would nowadays be regulated in their travels by a traffic robot made of recycled materials, designed by a woman engineer (Sasse & Haasen, 2018). The sounds of traffic in these places echo the relative values given to traffic diversity, access, regulation, speed, safety, privacy, and individuality in the place of recording. Traffic, cell phones, and Global Positioning System (GPS) triangulate and proliferate, underlining our desire for mobility.

Participants comment on how much more they hear road traffic during a soundwalk, realizing how much we must filter usually. Some can never escape its drone, those who live in the acoustic shadows of highways and major intersections. For example, one YouTube listener responded to a soundwalk recording, engaging in sustained conversation about living near the Decarie expressway in Montreal, and his attempts to live with traffic whose sounds filled and marked routines of daily life. The conversation with this listener also exceeded questions of frequency or amplitude. He spoke of what he heard in the traffic: diurnal, seasonal, and weather changes. He spoke of how the sound penetrated his apartment, suffusing the living space, especially in summer with windows open. He found the Balance-Unbalance soundwalk recording had a calming traffic sound, like rushing water. This listener later developed a meditation practice that hinged on mentally transforming traffic into rushing water, to make living with the traffic feel more natural (and inevitable, since the listener refers to nature as something that cannot be changed). Another soundwalker and sound artist, Victoria Fenner, says that she incorporates traffic into a meditational practice directly, while walking, by breathing along with particular vehicles as they pass her.

Some listeners are at first bored and frustrated by traffic sound and then search for texture, form, and meaning. A remarkable instance of mobile listening is related by composer and theorist (1992) James Tenney, experienced repetitively through the daily grind of driving the Holland Tunnel to get to the Bell labs in New Jersey in the early 1960s, when he was beginning his computer music work there:

> One day I found myself listening to these sounds, instead of trying to ignore them as usual … the sounds of the traffic became so interesting that the trip was no longer a thing to be dreaded and gotten through as quickly as possible. From then on, I actually looked forward to it as a source of new per-

ceptual insights. Gradually, I learned to hear these sounds more acutely, to follow the evolution of single elements within the total sonorous 'mass', to feel, kinesthetically, the characteristic rhythmic articulations of the various elements in combination, etc. Then I began to try to analyze the sounds, aurally, to estimate what their physical properties might be – drawing upon what I already knew of acoustics and the correlation of the physical and the subjective attributes of sound. (Tenney in Polansky, 1984, 154–5)

Tenney also notes the similarity of traffic sounds to ocean surf, echoing the ideas of the YouTube listener. But Tenney wanted to avoid the periodic nature of ocean sounds and his resultant "Analog #1: Noise Study" is aperiodic. He does say he believes there are some elements of ocean wave sound in this piece, but does not articulate what these are. Katherine Kline, psychoanalyst, noise musician, and scholar, thinks through the filter of psychoanalysis in her discussion of a soundwalk for the *Soundwalking Interactions* blog. Hearing "the far-away traffic hush" she speaks of psychic processes of enclosure and containment:

Here, the thought occurs to me that in a city, we are encircled by this sound, enmeshed in it, and in a sense *contained* by it. The perimeter of the city, overpasses and highways, form a sonic envelope of rubber on asphalt/moisture, a hushing, rising and fading shhh. This dull, sedating rhythm did some mysterious work on me.[1]

As a child I used to follow my mother around the house while she cleaned. I had a particular attachment to her vacuum cleaner, a source of warmth and noise. Wherever she might be vacuuming, I would lay my blanket down beside the machine and slip into its envelope. My mom and that machine and I were intermingled in a noisy aesthetic that connected us for a moment. Onto the next room she would go, and I would follow. My mother's machine not only cleaned the carpets, it held me through its mollifying sound.

We are held, as the psychoanalysts tell us, born helpless, gestating in an environment of objects, entities, capacities. We find ourselves wide-open, open-eared in this place … this sonorous bath points us precisely to a lack of closure, the opening-out of reception. Something is resounding within, as we're sounding-out. Our sonic envelopes have a way of expressing themselves through us. (Kline, 2013)

Like Kline's mum's vacuum cleaner, or the raver's woofer speakers, the broad noise band of traffic heard at a safe distance—and here perspective becomes very important—can become a continuous, soothing blanket that softens contact with the surrounding world, belying its origin in an ecotone, a landscape of tension; belying the terrifying howl and thump of air when a truck passes close by; belying the release of toxic emissions into the atmosphere; resonating; feeling like home. A search on "soothing diesel sounds" reveals engine recordings up to 12 hours long, meant as lullabies. Containment and enclosure can feel soothing and secure, or irritating and interrupting thought, depending on the context and the listener. And these feelings can change, depending on listening experiences.

The feeling of containment and comfort can come from physical distance from distant traffic, from mental distance during a soundwalk or composing process as one begins to reflect on and consider the sounds and how they interact, from aesthetic distance as traffic is framed and sweetened in various ways during processing of sound. A soundwalk frames traffic in time through a shared activity. A piece made from traffic sound, whether in situ or heard elsewhere, can frame the traffic and sweeten it through processing, filtering, harmonizing, rhythmicizing, aestheticizing the sound. Repeatedly in writing about urban traffic sound art projects, ideas of comfort and soothing emerge, of "cushioning the blow" as Noah Vawter describes it (Vawter, 2006, p. 16), of softening urban noise as the director of the Mass MOCA gallery writes in response to an artwork by Bruce Odland and Sam Auinger (Thompson in O & A, 2009, p. 65). Artists working in situ, such as Bruce Odland and Sam Auinger, or Gayle Young and Reinhard Reitzenstein, use tubing to filter and harmonize the traffic sound; in another work by Odland and Auinger, listeners can sit on speakers to feel sound through their bodies more clearly; mobile audio such as that by Noah Vawter (2006) or Samuel Thulin (Ng-Chan, 2012) plays with traffic sound rhythmically. Sound directs attention in different ways in these works that are structured through harmonizing, rhythmicity, vibration. Soothing persists.

I began the Montreal research projects with a sustained creative interest in traffic sound. As well as many recorded soundwalks that include traffic, several of my performance and installation works explore this subject. *Peace(π): A Slick Circus on a Discord of Flags* (1993) is a realization of the *Circus On…* score by John Cage in which (among other things) we used car parts such as a tail pipe and gas tank to perform improvised music about the first Gulf War; *In Transit* (1999), *Streetcar Turning* (1999), and *Streetcars Made for Saints* (2012) focus on sonic experiences of urban transit; *Canada Train Radio 1: Montreal to Winnipeg* (2003) is about long-distance train travel; *Toronto-Norwood-Toronto* (2008) is an interactive piece on long-distance cycling. Also, over the last decade I have worked on my home garden, planting extensive shrubbery to filter fumes and sound from the highway, and to provide habitat and shelter for birds and their song, a home landscaping/soundscaping project related to traffic.

For the Resonant Bodies symposium I made a new piece, *Spectral Traffic* (2013). My intention in making *Spectral Traffic* was to create a virtual conversation about traffic, integrating short samples of sound works by James Tenney, Victoria Fenner, and Katherine Kline based on their sonic traffic experiences, with various traffic recordings from soundwalks done through the *Soundwalking Interactions* project. I processed the sound, brushing through the frequency spectrum, focusing attention on the infrasonic rumblings that mark emissions of multiple diesel and gasoline engines, and the flights of tires, sirens, bells, and horns across the sound field. These sounds are juxtaposed with the joyful human environment of a bicycle race, recorded in the same location as some of the car traffic recordings, outside an apartment on Parc Avenue in Montreal. Finally, the piece explores the fantastic imaginary space of slowed experience, created by dwelling in and slowing the speed of the bicycle bell sound.

When I put a copy of the piece online, almost immediately there was a comment about how soothing the traffic sounded to one listener, and how the human sounds seemed to interrupt this soothing. There it was again. Perhaps the timing of sweeps through the spectrum, the fluidity of mix from one sound to another that makes the traffic interesting for me to listen to and interact with as a performer, strokes the sound into the listener's ear, making sounds that are periodic, feeling like breathing or ocean waves in their coming and going, rather than the aperiodic and less obviously rhythmic noise of dense road traffic. *Spectral Traffic* is the most abstract and processed of the traffic work so far, with its emphasis on frequency shifts. It attempts to explore traffic in many locations, taking disparate traffic sources, and mixing them together, never staying in one location long. Perhaps this level of abstraction led to a dream world that lost the tension of the everyday.

Listening to traffic must take place in time, since it relies on the transmission of sound. Deep understanding of traffic through a place relies on listening repeatedly for diurnal, weather-related, political, and other kinds of shifts in the sound environment, creating a more subtle and resilient sonic understanding of place. It is not possible to take an instant photo that will capture the scene. Soundscape seems too still in this case, missing the dimension of time; the more active word soundscaping, like landscaping, focuses on the importance of repeated attention and care through time to the understanding of a place, especially through listening. Allen Weiss, a keynote speaker at the Resonant Bodies: Landscapes of Tension symposium, later wrote: “The most complex landscape is the one most closely observed” (2015). Through these research-creation projects, everyday attention to guts as well as ears as we enter into different kinds of traffic, thinking about the ways that various technological antennae and ideas contribute to our listening, working with the sounds that we hear, and paying attention to how sound works are received, add to understanding about our participation in traffic.

Traffic suggests routes. It challenges binaries like country and city, freedom and containment, heard sound and felt sound, wilderness and urban living, traffic noise and rushing water, human sound and mechanical sound, soothing and irritation, focusing instead on how we get from one to another, what routes, vehicles, and practices we choose. We are traffic; the sound of traffic is the sound of our desire for mobility. But it remains hidden, masked. We quiet traffic in our imaginary creations, using bass roll-off filters and equalizers to push it further into the distance. Noise measurement protocols minimize traffic and other infrasonic industrial engines. Contained by traffic, we contain it in our imaginations. We can turn down the volume until the engine drone soothes us, meditate or dream with its sounds, aestheticize it harmonically or rhythmically. And traffic continues. … If we still listen, attention to the daily sonic experience of traffic can aid in thinking about how mobility is valued in different situations, and, in deeper ways, how traffic moves all of us.

Note

1. Katherine Kline’s February 12, 2013, Montreal Soundwalk referred to in this passage can be found on Soundcloud: https://soundcloud.com/katherinekline/feb-12-2013-soundwalk-montreal

References

Bijsterveld, K. (2008). *Mechanical Sound: Technology, Culture, and Public Problems of Noise in the Twentieth Century.* Cambridge, MA: MIT press.

Garon-Sayegh, P. (2018). Noise Regulation and the ‘A’ Scale. Legal Report for A. McCartney.

ICI Berlin. (2013, June). *Resonant Bodies: Landscapes of Acoustic Tension* [Conference Schedule]. Retrieved from https://www.ici-berlin.org/wp-content/uploads/2017/03/Symposium-Resonant-Bodies_Programme.pdf

Kline, K. (2013). In Search of the Vacuum Cleaner: February 12 Soundwalk, Montreal. *Soundwalking Interactions*. Retrieved June 10, 2013 from https://soundwalkinginteractions.wordpress.com/2013/03/15/in-search-of-the-vacuum-cleaner-february-12-soundwalk-montreal/

Ng-Chan, T. (2012). *Detours: Poetics of the City: City Ditties*. [Online] Agencetopo.qc.ca. Available at: http://www.agencetopo.qc.ca/detours/cityditties_en.html. Accessed 28 Oct 2018.

O+A (Odland, B., & Auinger, S.). (2009). Reflections on the Sonic Commons. *Leonardo Music Journal, 19*, 63–68.

Polansky, L. (1984). The Early Works of James Tenney. *Soundings, 13*, 154–155. Santa Fe, NM.

Sasse, E., & Haasen, B. (Producers). (2018). *Digital Africa Traffic Robot* [Video]. Retrieved July 27, 2018 from https://www.arte.tv/en/videos/077470-006-A/digital-africa-6-6/.

St. Pierre, R. L., & Maguire, D. J. (2004). The Impact of A-Weighting Sound Pressure Level Measurements During the Evaluation of Noise Exposure. Paper Presented at *Noise-Con*, INCE-USA, pp. 1–8.

Tenney, J. (1992). *Meta + Hodos: A Phenomenology of Twentieth-Century Musical Materials and an Approach to the Study of Form: And META Meta Hodos*. Hanover, NH: Frog Peak Music.

Vawter, N. (2006). *Ambient Addition: How to Turn Urban Noise Into Music*. MA Thesis, Massachusetts Institute of Technology (MIT), Cambridge, MA, Department of Architecture. http://hdl.handle.net/1721.1/36161

Weiss, A. S. (2015, May 27). Manifesto for the Future of Landscape. *Writers in Kyoto*. Retrieved June 20, 2018 from http://www.writersinkyoto.com/2015/05/weisss-manifesto-for-the-future-of-landscape/

Wilson, A. (1998). *The Culture of Nature: North American Landscape from Disney to the Exxon Valdez*. Toronto, Canada: Between the Lines.

CHAPTER 13

Acouscenic Listening

Sean Taylor and Mikael Fernström (Softday)

Introduction

In this chapter we introduce the concepts of acouscenic listening and creative soundwalking. We argue that this is a sonically engaged arts practice that can help democratise the relation between artists, performers and communities of interest. The term "acouscenic listening" is a portmanteau of the words *acoustic* and *scenic* (as in picturesque), which is the concept of experiencing constituent sonic events characterising a landscape. All our human senses of perception seem to be at optimum when walking. Part of a soundwalking activity is to stop and listen to sounds that may be particularly interesting. It appears to be a natural human ability, and we assume that this is a result of how humans have evolved since the palaeolithic hunter-gatherers. It can also be said that everyone who travels has a story to tell; hence a soundwalk can have an implicit narrative structure. What makes a soundwalk creative is the collective making of a soundmap that can be used as a score for performance.

S. Taylor (✉)
Limerick Institute of Technology, Limerick, Ireland
e-mail: sean.taylor@lit.ie

M. Fernström (Softday) (✉)
University of Limerick, Limerick, Ireland
e-mail: mikael.fernstrom@ul.ie

M. Droumeva, R. Jordan (eds.), *Sound, Media, Ecology*,
Palgrave Studies in Audio-Visual Culture,
https://doi.org/10.1007/978-3-030-16569-7_13

Background

Acouscenic listening draws upon several methodologies, ranging from acoustic ecology to social art practice, laying a foundation for a creative turn in our approach to creative soundwalking, which endeavours to be a unique methodology in sound and performance art, encouraging greater interaction between artists and audience in creating, educating, performing and recording sound/body-based work. The focus of Softday's acouscenic listening and creative soundwalk practice cultivates an understanding and appreciation of soundscape to body on a finely tuned level, expanding the potential for connection and interaction with one's environment, body, technology and performance together with others, in sound and related arts. When working with the quotidian this is realised through focused soundscape walking meditations, a methodology that has become a key element in our acouscenic listening practice. The following sections outline the main theories and practices that acouscenic listening and creative soundwalks are based on.

Listening

In 1974 the American composer Pauline Oliveros published *Sonic Meditations*, a seminal work that broke away from the prevailing traditions of Western music, and challenged the need for standard music notation in favour of prose instructions (similar to Alan Kaprow's Happenings or Fluxus event scores). In 1980 her research led her to establish an aesthetic philosophy she outlined as Deep Listening, which redefined listening as an art in itself, and in 1985 she established the Deep Listening Institute in Kingston, New York. Oliveros (2005) defined Deep Listening as:

> a practice that is intended to heighten and expand consciousness of sound in as many dimensions of awareness and attentional dynamics as humanly possible. The source for Deep Listening as a practice comes from my background and experience as a composer of concert music, as a performer and improviser. Deep Listening comes from noticing my listening or listening to my listening and discerning the effects on my bodymind continuum, from listening to others, to art and to life.

Oliveros outlined the need for a fundamental change in the way that we all listen, how listening (an active and attentional process) is different from hearing (a passive and involuntary physical activity), and how the act of habitual listening may be a catalyst for cultivated reception and creative actions. Oliveros defined quantum listening as "listening in as many ways as possible simultaneously – changing and being changed by the listening" (2010, p. 74).

In 2009 Sean Taylor began working with Pauline Oliveros at the Deep Listening Institute in New York. Whilst completing a Deep Listening Certificate programme at the Deep Listening Institute he understood how sound from a creative, cultural and cognitive perspective mediates one's relation with both space and place. Oliveros outlined a rationale for the *listening effect*, or how what we hear may be processed. Quantum listening directs our attention in a non-judgemental way to what is heard, amassing meaning, and interpreting the meaning of what has been listened to, while deciding on further creative actions. The understanding and application of the quantum listening methodology constitutes a key starting point towards Softday's ongoing investigation of acouscenic listening practice.

Walking

As an exploration of a soundscape of a given area, soundwalks have become an important approach for researchers in acoustic ecology and sound art practices. Soundwalking, originally proposed by the acoustic ecologist and composer R. Murray Schafer, emerged within acoustic ecology research as an active form of exploration of soundscapes (Westerkamp, 1974, 2007). The skill of listening is individual and subjective, influencing how we listen to the world. The skill set required to listen attentively to sounds in a soundscape and to further extrapolate those sounds for creative development is dependent upon training an individual's ability to alter and focus their listening habits. The appreciation of acouscenic listening as a viable art practice also relies on the creative practitioner's ability to unlock potential narratives of soundscapes.

Westerkamp (1974) suggested that a participatory soundwalk can have three specific functions: orientation, dialogue and soundwalk composition. As both an environmental activist and a composer, Westerkamp advanced the concept that soundwalks as a form of dialogical exchange with the acoustic environment could be intrinsically linked to politics, embodiment and storytelling traditions. Traditional soundwalking meditations can be either solitary or shared experiences, encouraging participants to create subjective maps based on areas of appeal, mapping mentally and mindfully a relationship to place and memory through environmental sound. Softday's creative soundwalk differs from the traditional soundwalk in terms of its objectives to locate the practice within the everyday, to encourage its participants to be active listeners, researchers and creative participants. The creative soundwalk straddles a polemic between epistemological and practical realms in order to establish a creative space for meaningful dialogue between the listener and the everyday.

The Everyday

The acouscenic listening approach to creative soundscape mapping can reveal a myriad of sonic events that are often mundane, habitual or ambiguous. Everyday language also generates sonic insights, directing our listening, outlining and obscuring it, but most importantly it also serves to unlock our ears. A desire to reveal the minute constituent parts of a soundscape finds purchase in the writings of Georges Perec (1997), who along with Michel de Certeau and others integrated new ways of engaging with urban spaces through concrete experiences of the everyday. Perec asks:

> How should we take account of, question, describe what happens every day and recurs everyday: the banal, the quotidian, the obvious, the common, the ordinary, the infra-ordinary, the background noise, the habitual? (1997, p. 210)

Perec employed neologisms, like the term *infra-ordinary,* to designate an excessive type of attention to the everyday, and *endotic* to communicate an everyday that is neither routine nor exceptional. Perec's task therefore was to foreground those elements that are essential to our lives, as they are frequently overlooked when traditional notions of meaning are employed. Perec directs us to make inventories of minute details, because the everyday shapes our identity and offers invaluable insights into our own anthropology.

A similar methodological approach is employed on the creative soundwalk where participants are encouraged to consider and map the *infra-ordinary* sounds that appear both essential and trivial as divulging elements of an everyday sonic narrative of a soundscape. The principal aim of our research is to reveal the creative process that integrates everyday *endotic* sounds with sound art and music composition. The process draws from a historical auditory line of enquiry instigated by Thomas Edison, Luigi Russolo through to John Cage and Cornelius Cardew (among others), seeking to insert the sounds of the banal and the everyday directly into live performance. The underpinning for acouscenic listening as a sound art practice comes from Softday's collaborative experience as social art practitioners, composers of acousmatic music and as performance artists and improvisers. It also draws upon frameworks that help guide where and how the practice may be positioned within the everyday. These include the practices of composers and educators such as Schafer, Westerkamp, Oliveros, the essayist Georges Perec, the philosopher Henri Lefebvre and art movements such as Dada, Fluxus and the Situationists.

Walking, the most natural human behaviour for exploring our environment, is also important as the walk is simultaneously the source material out of which art may be produced and also the method of the artistic transaction. In an urban context, the city soundscape offers a myriad of possibilities for the chance encounter or improvisation to happen. The action of the soundwalk through a soundscape can be challenging, therapeutic and transformative for the participant due to the constant interaction of all the sense modalities, an experience that Bachelard (1994) refers to as "a polyphony of the senses", the ear collaborating with the body and the other senses.

Our approach to creative soundwalking as a way of engaging with a city or other spaces may be considered closer to the *dérive* or *drift*, defined by Guy Debord (1958) and the Situationists as "a technique of rapid passage through varied ambiences. *Dérivés* involve playful-constructive behaviour and awareness of psychogeographical effects, and are thus quite different from the classic notions of journey or stroll". We differentiate between the conditions of the *dérivé* and the *flâneur* as defined by both Charles Baudelaire (1965) and Walter Benjamin (2002), in that we consider the *flâneur* to be a passive wanderer of the city, not an active participant in the actions of the city.

Acouscenic listening, through its application of the creative soundwalk, promotes the value of non-judgemental listening and actively seeks to deconstruct the boundaries between the listener and the listened to. Non-judgemental listening attempts to break down artificial impediments between listener as subject and listened to as object, an interpenetration that may produce a mind of harmony and attainment of creativity. This is similar to a perceptual condition that anthropologists refer to as *paralinguistic signals*—those gestures and inflections that lie beyond, underneath and beside language.

Socially Engaged

The creative soundwalk encourages interaction of the individual listener to space or place through immersive or embodied experiential mapping and a basic understanding of psychogeography. In order to do this, two elements are applied. The first is the *social turn*, a term used in 2006 to describe the return to a socially engaged art form that is collaborative, participatory and involves people as the medium or material of the work. When applied to the creative soundwalk this approach to social art practice may be referred to as *intermedial*, a concept that challenges and "re-calibrates inherited understandings of what is within and what is without the art event" (Jackson, 2011). This concept is similar to artist Dick Higgins' (1965) description of how Fluxus artists worked across or between creative genres, an approach that he defined as intermedia. An intermedial approach to the creation of significant artworks has been at the heart of our acouscenic listening practice since 2012.

The second integral element is the *creative turn*, which specifically refers to the process by which the co-authored subjective maps created by participants and the artist on a creative soundwalk are constructed. What we understand by *creative turn* in this context refers to a conscious moment within the process of a creative soundwalk where the original topographical information is put to one side, and new thinking about the potentiality of performing the map as an improvised graphic musical score emerges. The word *potential* is important in this context as not all creative soundwalks we have undertaken with the public have led to creative sound works. By potentiality, we mean the *possibility* to act upon mapped information, which is in itself not limited by the mapping process. We contend that the creative soundwalk creates moments of speculation, experimentation and reflexivity, constructing spaces of experimentation and improvisation without the constant demand of proven results. Creative outcomes from a creative soundwalk perspective are dependent upon consensus within any group of participants as to whether or not they wish to take the topographical information collected on the soundwalk beyond just a graphical artefact. Sometimes the reality of time constraints within the workshop delivery also has an impact on creative delivery.

The soundscape of a city is more than just a cacophony of background noises and rhythms with no clear purpose or value. Lefebvre implies that the cyclical daily rhythm of life conveys meaning significant to the disclosure of a city's purpose: "once one discerns relations of force in social relations and relations of alliance, one perceives their link with rhythm" (Lefebvre, 2013). Sound textures refer to the strata of sounds within a (musical) composition and their relationships to each other. The aural texture of an urban soundscape may be constructed by an attentive listener sensitive to the combinations of competing sound signals arising from background noise. These aural textures may be considered as static images of the auditory system (Rabinowitz & King, 2011).

What Happens on a Creative Soundwalk

While the basic soundwalk can be exploratory, scientific, experiential and so on, the creative soundwalk is predominantly undertaken in collaboration with specific communities of interest (Lennon & Taylor, 2012). We define the creative soundwalk as a structured immersive excursion whose main purpose is to engage participants to become both active listeners and creative sound makers.

A typical workshop begins with an introduction by the artist to quantum listening, psychogeography, soundscape practice and the creative soundwalk as a methodology to creatively transform acoustic space through the creation of collaborative and improvisatory sonic art works. Participants are also introduced to R. Murray Schafer's (1977) three main elements of the soundscape: keynote sounds, sound signals and soundmarks, which are important aspects of any unique space. Participants are also introduced to *Qigong* and *Tai Chi* exercises in order to consciously quieten the mind and body and bring listening attention to the soundscape. Participants undertake adapted ear cleaning exercises, which were first proposed by R. Murray Schafer (1967), who emphasised that ears should be metaphorically *cleaned* through specialised listening practices as a prerequisite to deeper listening. It is also important to take time to help the listening participant to learn a process of aural wayfinding. Auditory engagement with space is the result of a reciprocal process between the listener and the sonic environment. Before a creative soundwalk starts, the aims and processes are contextualised and communicated to all participants.

The facilitator identifies a route for auditory exploration, and provides a map of the creative soundwalk route for each participant. The walk may be one to two hours in duration. The map contains some key listening points on the route for longer more immersive listening experiences. On a creative soundwalk traditional field recording and soundscape composition methodologies can be put aside in favour of a collective and diaristic approach to registering sound events in preparation for a subsequent performative aspect. A field recordist aiming to document and preserve a soundscape might focus on the technicalities of the recording process, while someone working on a soundscape composition might attempt to capture the emotional quality of sounds in the moment. The creative soundwalk participant, on the other hand, is asked to focus on listening with a cartographic sensibility, to notice as many sound events as possible as they pass through various soundscapes so that they can document their experiences of the soundscapes visited, combining quantum listening and note taking to generate an alternative experiential map of the excursion. This map, in turn, will be combined with the maps of the other participants to generate a collective map that can then serve as the basis for performance, adapting the role of the soundwalk from solitary listening or recording in the moment to a rehearsal of space for subsequent creative engagement on a collective level.

The creative soundwalk is normally undertaken in silence. Walking in silence is an unfamiliar experience for many, but it is an important factor that creates the space where the main aim of the creative soundwalk is to heighten the participant's concentration to the myriad sounds and other sensory dynamics of the moment and environment being walked through. In this engaged state, the mediating participant is aware of all that happens with real transient and situated sound of place occurring within a real geographical time frame. The soundwalker engaging with acoustic space creates a scenario, an improvisational interrupt, a change of perspective that deepens the embodied listening experience, where each listener is free to interpret and contribute to a fluid sonic environment at any given moment.

Upon completion of the creative soundwalk a soundmap of the walk is collectively co-authored. Referring to their notes, each participant registers sound events from the walk. Participants' annotations are spatially distributed as the experienced space is remembered, using colour coding to signify, for example, biophonic, geophonic and anthrophonic sound sources. The soundmap produced by participants on a creative soundwalk is ultimately representative (and artwork in itself), which suggests no limitations as to how it may be represented or remediated (see Figs. 13.1 and 13.2).

A creative turn is then applied to the completed soundmap, with agreement of all participants. This involves treating the soundmap as a graphic score for possible creative and performative outcomes. Preparatory work for the performance of the graphic score involves group Qigong/Tai Chi

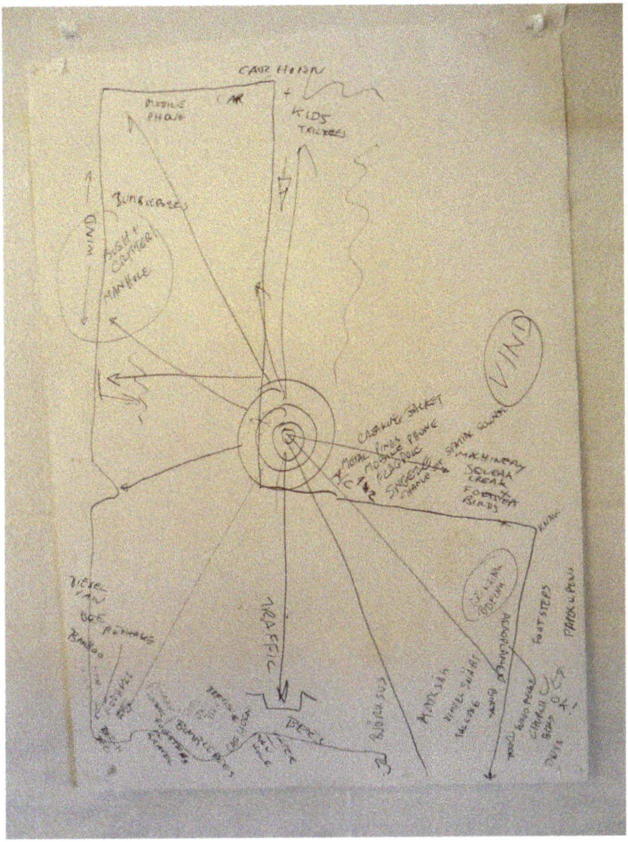

Fig. 13.1 Soundmap of the village of Harplinge, Sweden, 15th of April 2014

and improvised sonic meditation exercises, which are led by the facilitator. Participants are encouraged to use their own bodies and voices as sounding instruments in order to perform the graphic score. The concept of *sounding* as opposed to *singing* a graphic score is introduced as it allows for a more democratic performance of the work. After further rehearsals, the visual literal reading of the score is abandoned in favour of a more improvised rendition that incorporates movement elements and listening to what other performers are sounding. These creative experiences and interactions are similar to what de Certeau (1984) describes as a "spatial acting-out of place".

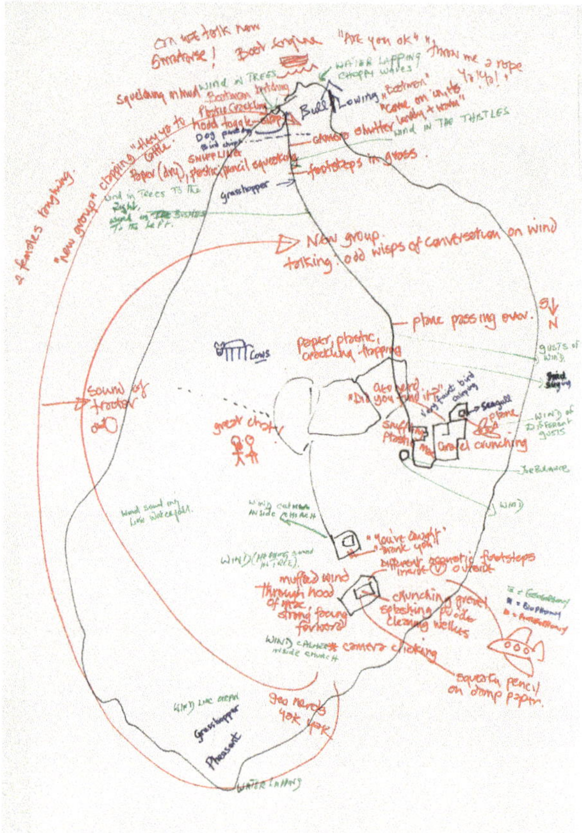

Fig. 13.2 Soundmap of Holy Island, Lough Derg, Co. Clare, Ireland, 20th of August 2016

This approach significantly changes the creative nature of the performance and opens up a potentially more meaningful and emotive rendition of the score. This creates the possibility for a multisensory experience of everyday interactions with both space and place and potentially challenges aesthetic distinctions between high art and popular culture. We are therefore able to re-evaluate the conditions that inform our collaborative and co-authored sound art practice and address the thorny question of the function of the aesthetic of the non-professional or amateur in our work. We prefer to use the French translation of the origin of the word amateur meaning *lover of*. When we speak of an aesthetic of the amateur, we also refer to a Cageian definition of our relationship to sound. Cage's (1961) legacy is extensively associated with the idea that any sound, whether it is part of a formal musical performance or not, may be listened to musically.

Consequently, a creative aesthetic is negotiated and agreed among the participants, as the development of the work progresses. This is why the concept of sounding rather than singing the graphic score allows for a more democratic performance of the work. The conscious listening and performance training leads to awareness that creativity is disclosed in moments. Being fully present in these moments fosters an understanding of the transformative possibility of creative encounters, instituting a mode of listening that French philosopher Jean Luc Nancy designates as "the homology of sound to self as a structure of referral" (Kane, 2012).

From a creative perspective, it is interesting to note that a first iteration of a soundmap composition usually results in the performers attempting to literally imitate the mapped sounds of the environment from the graphic score (see Fig. 13.3). After further work, the literal visual reading of the score is abandoned in favour of a more improvised rendition that incorporates movement elements whilst listening to what the other performers are sounding. This approach significantly changes the nature of the performance and facilitates a more empathic interpretation of the score. At each performance iteration the score can be recorded and played back to the participants for further discussion on the aesthetic of the work. In this respect, the composition is always in a state of flux or incompletion. Both the co-authored soundmap/graphic score and audio

Fig. 13.3 Generating creative soundmap and working on live improvisation based on soundwalk, 20th of August 2016

recordings can then be shared between all participants with agreement that participants can do additional work on the graphic score or the recordings if they so wish.

The creative soundwalk builds on the work of the French Marxist philosopher Henri Lefebvre's concept of spatiality and critical thinking about urban space and everyday life in relationship to body, action, and the constructed environment. Lefebvre maintained that space is fabricated within a dialectic relationship between social action and spatialisation. He suggested that space is a social construction, and not a complete or naturally occurring phenomenon. Lefebvre (1991) declared that space is an infinitely contested and politicised arena. Lefebvre argues that space is never passive, neutral or a pre-existing given, but rather it is an ongoing production of spatial relations. He states that there are

> "an indefinite multitude of spaces, each one piled upon, or perhaps contained within the next" and that "spatial practice consists in a projection onto a spatial field of all aspects, elements and moments of social practice"(1991, p. 8).

He also claimed that "(Social) space is a (social) product" and that these types of spaces embody a myriad of social relationships. Social spaces are not abstractly constructed; rather they are socially and collectively produced, and therefore the practice of acouscenic listening reveals an understanding of how such social spaces may be constructed and utilised for creative purposes. This connection of space to the everyday was adapted by a number of European avant-garde artist groups, in particular the Situationist International movement (1957–1972) led by Guy Debord. In his 1959 essay "Towards a Situationist International", Debord stated that an individual's life may be considered as "a sequence of chance situations", which if experienced in isolation acts as a social restraint and encourages passive spectatorship. Debord argued for the creation of collective situations that are actively constructed by spectators as a form of *unitary urbanism*, thus eliciting the spectator's "capacity to revolutionise his own life" (Debord, 2006). In the late 1950s, the Situationists originated a number of creative urban investigations, which included the *dérive* (drift) and psychogeography (the psychological effect of urban environments on the individual).

Lefebvre outlined what he classified as "a conceptual triad" in explaining how spaces may be produced. We have revised this triad in order to explain how these categorisations may be applied to the practice of acouscenic listening:

- *Spatial practice* makes reference to the production and reproduction of spatial relations between objects and *products*. In the practice of acouscenic listening this may be equated to the creative production and remediation of spatial relations between sound events in a soundscape as experienced by an attentive listener and the representation or remediation of this information as sound art.
- *Representations of space* refer to conceptualised space, the space of architects, scientists, acoustic ecologists and sound artists, or those who identify with what is lived and listened to with what is creatively conceived.
- *Representational spaces* are everyday experiences where meaning emerges as a result of a dialectical relation between acouscenic listening practitioners and the creative remediation of space.

Lefebvre (1991) reminds us that "in seeking to understand the three moments of social space, it may help to consider the body. All the more so in as much as the relationship to space of a "subject" who is a member of a group or society implies his relationship to his own body and vice versa". Lefebvre's concepts of everyday life and urban space also resonate with the Deep Listening work of Pauline Oliveros and the soundscape compositions of Hildergard Westerkamp, who both locate the listening body at the centre of situated soundscapes. Sound can be an invasive phenomenon of everyday experience in that it assists our engagement with, immersion in, and commentaries with the environment in which we live.

Softday uses components of both Lefebvre's and Debord's spatial theories, in abstraction from their complete work, in order to illustrate our interest in temporal, social, cultural and everyday concerns within the practice of acouscenic listening. In its attempt to promote a creative attitude towards soundscape analyses, the acouscenic listening approach to the use of the creative soundwalk as a method of artistic intervention similarly questions (amongst other issues) production and authorship issues with regard to how urban and rural spaces may be creatively reconstructed. The creative soundwalk as a collaborative *dérive* involves a primary determination of soundscapes, unlocking the possibility for a subjective understanding of recording, mapping and applying form and value to the "inherent indeterminacy" of the everyday (Pable & Hutton, 2015). The desire to reveal the language of the constituent parts of a soundscape can be seen in the writings of Michel de Certeau (1984) and Georges Perec (1997), who integrated new ways of engaging with urban spaces through concrete experiences of the everyday. de Certeau puts it succinctly: "the approach to culture begins when the ordinary man *becomes* the narrator, when it is he who defines the (common) place of discourse and the (anonymous) space of its development" (de Certeau,1884, p. 5). Perec asks: "How do we speak of these common things, how to track them down rather flush them out ... how to give them meaning, a tongue, to let them, finally, speak of what is, of what we are" (Perec, 1997, p. 210).

From an acouscenic listening perspective the creative soundwalk encourages the interaction of the individual listener and space through immersive experiential mapping and a basic understanding of psychogeography. In John Drever's (2009) assessment that "the salient concern in soundwalking is everyday life" is a key concern of a creative soundwalk in the corporeal exploration of sound, location, narrative and its relation to the everyday. Punctilious interaction with soundscapes of the everyday implies a shared aesthetic experience by both artist and participants, in order to validate listening behaviours, aural perception of the soundscape traversed and the creative interpretation of the mapping of the soundscapes. The creative soundwalk may also be considered as a form of liminal activity where the participant's own embodied physicality engenders a temporal dialogue between active listener and the aural architecture of place within the everyday. Passing through diverse aurally textured soundscapes, the narrative of place/space is gradually revealed to the active listener, and the overdependency by participants on an ocular interpretation of these sonic environments gradually diminishes.

This enables the listener to become fully immersed in the present at all times, both during the soundwalk and the improvised performance of the graphic score. Being continuously in the moment enables the participant to let go of preconceived views, and enter into what can be termed a state of flow, "the state in which people are so involved in an activity that nothing else seems to matter" (Csikszentmihályi, 1990). Improvisation implies a temporal space of "not-knowing", where the outcome cannot be defined in advance, but rather dialogically and creatively unfolds over time through collectively doing, reflection and editing. The process and outcome of a creative soundwalk are not invalidated by an initial lack of skill, but can become even more thought-provoking as a result of actions that are not fully comprehended. The process is one of trial and error, trying out what works more effectively in a given arrangement based on the sonic information and material (in most cases the human voice).

We are not unduly concerned about how to formally analyse a particular sound composition that creates and describes itself in the genesis of its sounding. The question that is of interest to us is (in the absence of a set of rules to guide a collaborative sound work) how does the interaction of the participants/performers and information contained in the soundmap shape the direction of the composition/performance. It has been our experience that the more open the arrangement that facilitates this space of not-knowing, the more active the interaction between the performers becomes. An inherent aspect of an acouscenic listening approach to creative soundwalk practice is playfulness, an essential ingredient that we believe is fundamentally suited to encourage creative self-expression for the untrained participant. Play liberates the listener from an overdependency upon competition, cogent discourse, rationality, and aesthetic angst, in relation to music/sound inherent through pre-existing controlling paradigms of social and cultural conditioning. In their critique of the primitive social functions of play the Situationists stated that "its goal must be at the very least to provoke conditions favourable to direct living" (Situationniste #1, 1958).

Remediation

From an environmental or ecological perspective remediation is understood as the process of improving a situation or correcting a problem, whereas a creative remediation operates in a space where one medium is remade in another. In order to remediate our work we identify a problem and begin to look for ways to represent data associated with the problem in order to reform it. A typical Softday remediation project addresses this dualistic approach across a multiplicity of phases. Our remediated approach considers concept, research, planning, assessment of counteractive alternatives for creation of new space, through to construction and implementation of the chosen remedy as *secondary/new* artwork or performance. Any performance thereafter is also a remediation, in an allegorical sense. When we further remediate a work, at locations different to the original setting and without the original collaborators, we become an alternative community of sonic "narrators and translators" to the original community of interest we worked with. This *new work* or *new space* is also located in a state of flux as we endeavour to re-represent it out of its original context, seeking to communicate the deeper affective currents and contestations that were contained in the original work.

Summary

An acouscenic listening and creative soundwalk approach to working with the public may democratise the relationship between the practitioners and a community of interest. The sound artist's role in this process is both socially communicative and creatively pedagogic, working with participants to share expert and lay knowledge and allowing participants to find their voice or form of expression that can coexist with others in a communal discourse. The development of acouscenic listening as a *sonically engaged* practice remains in a continual state of flux. As a mode of thought and an original practice of being, it does not exist outside of its connective dialogues, collaborative participations and creative happenings, but retains its capacity to be, incarnate, innovative and endlessly transformative.

References

Bachelard, G. (1994). *The Poetics of Space*. Boston, MA: Beacon Press.

Baudelaire, C. (1965). *The Painter of Modern Life and Other Essays*. London: Phaidon, Orig: Published in *Le Figaro* in 1863.

Benjiman, W. (2002). *The Arcades Project* (pp. 416–455). Cambridge, MA: Harvard University Press.

Cage, J. (1961). *Silence*. Middletown, CT: Wesleyan University Press.

Csikszentmihályi, M. (1990). *Flow: The Psychology of Optimal Experience*. New York: Harper-Collins Publ. Inc.

de Certeau, M. (1984). *The Practice of Everyday Life*. Berkeley, CA: University of California Press.

Debord, G. (1958). Definitions. Paris: Internationale Situationniste, (1).

Debord, G. (2006). *Guy Debord Works* (A. Debord & J-L. Rançon, Eds.). Paris: Gallimard.

Drever, J. L. (2009). Soundwalking: Aural Excursions into the Everyday. In J. Saunders (Ed.), *The Ashgate Research Companion to Experimental Music* (pp. 163–192). Aldershot, UK: Ashgate.

Jackson, S. (2011). *Social Works – Performing Art, Supporting Publics* (pp. 27–29). Oxford, UK: Routledge.

Kane, B. (2012). Jean-Luc Nancy and the Listening Subject. *Contemporary Music Review, 31*(5–6), 439–447.

Lefebvre, H. (1991). *The Production of Space* (pp. 8–40). Oxford, UK/Cambridge, MA: Wiley-Blackwell.

Lefebvre, H. (2013). *Rhythmanalysis: Space, Time and Everyday Life* (pp. 67–75). New York: Bloomsbury Academic.

Lennon, M., & Taylor, S. (2012). Straddling the Polemic. Limerick. http://www.lit.ie/lsad/Assets/Praxis%20MASPACE%20PDF.pdf

n.a. (1958). *Contribution to a Situationist Definition of Play* (Vol. 1). Paris: Internationale Situationniste.

Oliveros, P. (2005). *Deep Listening: A Composers Sound Practice* (pp. 23–24). New York: iUniverse, Inc.

Oliveros, P. (2010). *Sounding the Margins: Collected Writings 1992–2009*(p. 74) (H. Lawton, Ed.) Kingston, NY: Deep Listening Publications.

Pable, A., & Hutton, C. (2015). *Signs, Meaning and Experience: Integrational Approaches to Linguistics and Semiotics* (1st ed.). Boston: De Gruyter Mouton.

Perec, G. (1997). *Species of Spaces and Other Pieces* (p. 210). London: Penguin Books.

Rabinowitz, N., & King, A. (2011). Auditory Perception: Hearing and Texture of Sounds. *Current Biology, 21*(23). Science Direct. Retrieved from http://www.sciencedirect.com/science/article/pii/S0960982211011882

Schafer, R. M. (1967). *Ear Cleaning – Notes for an Experimental Music Course.* Toronto, Canada: BMI.

Schafer, R. M. (1977). *The Tuning of the World.* New York, NY: Knopf. Reprinted as *The Soundscape: Our Sonic Environment and the Tuning of the World.* Rochester, VT: Destiny Books, 1994.

Westerkamp, H. (1974). Soundwalking. *Sound Heritage, 3*(4), 18–27.

Westerkamp, H. (2007). Soundwalking. In A. Carlyle & I. Noble (Eds.), *No Autumn Leaves: Sound and the Environment in Artistic Practice* (p. 49). Paris: Association Double-Entendre. Retrieved from: http://www.sfu.ca/~westerka/writings%20page/articles&20pages/soundwalking.htm

CHAPTER 14

Evening of Sounds: Auditory Cultures in Radio Call-in Programmes

Heikki Uimonen

Introduction

Over the last six years, the Finnish Broadcasting Company has been airing a call-in programme on environmental sounds. A three-hour live programme called *Äänien ilta* (Evening of Sounds) has approximately 220,000 listeners representing 3.7 per cent of the country's population of 5.4 million (Hannula, 2016). The programme consists of sound requests and callers' narratives recounting personal experiences and memories of contemporary and historical sonic phenomena from urban, agricultural and natural settings. Sound requests are received via email and over the telephone and fulfilled with the help of sound archives and present-day soundscape recordings to present content familiar to listeners from their environments. The programme is founded on the basic principle of the radio: to broadcast sound content which is emotionally and intellectually meaningful to listeners. Each show features two or three visiting scholars representing environmental sound research, sound art and sound design to help unpack the language of sound opened up by the listeners themselves.

H. Uimonen (✉)
University of Eastern Finland, Joensuu, Finland
e-mail: heikki.uimonen@uef.fi

M. Droumeva, R. Jordan (eds.), *Sound, Media, Ecology*,
Palgrave Studies in Audio-Visual Culture,
https://doi.org/10.1007/978-3-030-16569-7_14

Evening of Sounds offers a platform for discussion and reflections on environmental sounds, which are produced and interpreted under specific historical conditions, thus bringing forth personally and culturally meaningful sonic phenomena. The frequently requested agricultural and, in some cases, disappearing and vanished sounds are accompanied by more contemporary urban and media soundscapes, such as historical time signals, jingles and the voices of popular radio play characters. A unique feature of the programme is that the audience and the hosts are listening to sounds and soundscapes attentively, a different approach from what one usually takes in everyday surroundings, inviting active personal and communal engagement with material that might often go unattended. This sharing of collective and personal sound memories and narratives sustains sociocultural history on sound but also feeds back by evoking and constructing new memories for the listeners, changing their experience of everyday sounds in their lives.

This chapter studies the cultural, social and aesthetic nature of the relationship between the kinds of sounds requested on the *Evening Sounds* programme and considers how their functions and meanings shift in this new media context. I hypothesise that this programme functions in parallel to those of music call-ins; the shared community aspect of the requests necessitates the fellow listeners' soundscape competence, referring to the ability to interpret sounds of a common cultural background. The programme will thus be theoretically interpreted and critically evaluated in relation to the idea of the "radiogenic," referring to experiments made specifically for the medium of radio. The 1920s sound and radio art of Dziga Vertov and Paul Hindemith will provide historical context for the *Evening of Sounds* programme.

Evening of Sounds stems partly from the pedagogical goals of soundscape studies in its attempt to make radio audiences listen to their environment attentively. At the same time, the programme democratises scientific research and is in dialogue with the audience: the feedback has evoked new research questions and action points among the scholars. The programme also raises questions about whether or not public service radio should be responsible for taking care of the sonic cultural heritage by preserving, creating and maintaining it in the context of contemporary transformations of the soundscape and, if so, how that might be carried out. To that end, this chapter elaborates on the role of radio in facilitating dialogue, preservation and cultural memory-making with respect to environmental sounds.[1]

Radiogenic Programmes

Since January 2012 Yleisradio (Finnish Broadcasting Company, FBC) has broadcast ten *Evening of Sounds* radio programmes. Three-hour call-in programmes are broadcast on Sunday evenings with approximately 220 000 listeners. Listeners are encouraged to request environmental sounds and share their memories and experiences with the radio audience and studio personnel. Over 800 requests have ranged from a grandfather's clock via shingle machine to the sound of effervescent vitamin tablets. Different themes of the programmes reflect broader social soundscapes, individual sound sources and research projects revolving around changing sonic environments with simple titles such as "Amazing World of Sounds," "Sound Stories," "Everyday Sounds," "Frightening Sounds," "Machines and Contraptions," "Transforming Finnish Soundscapes," "Sounds from the Pockets" and "Sounds of Tourism."

In a historical context, the programme content can be evaluated with the help of the radiogenic discourse. The BBC refers to radiogenic shows as "programmes made specifically for the medium of radio utilising to the maximum its distinctive qualities" (Chignell, 2009, p. 93) or that display an optimum aesthetic use of sound (Lacey, 2013, p. 93). The concept of "radiogenic" is related to factors that are present in the context of recording and broadcasting sounds. To some extent it is analogous to the concept of the photogenic, which refers to poetic, uncanny and sublime properties of a subject captured by photography, properties that highlight the meeting of artistry with a particular media form. "Radiogenic" refers to aspects that are "only evident in the recording and broadcasting of sound and that reveal or express an encounter with some sort of truth." Radio's "artificial reality" is constructed in space, subjectively perceived or reconstructed by individual listeners, while at the same time having some objective reality by its simultaneous consumption by a crowd of individual listeners (Lacey, 2013, pp. 93–94). The technical construction of artificial reality is embedded in Michel Chion's concept of "phonogeny" (1994, pp. 101–103), referring to the propensity of given voices to have such qualitative aspects that when they are recorded and played they can make up for the sound's real source by means of another kind of presence specific to medium. Technically this means that some voices have timbres more favourable for technology than others. The concept underlines the processual nature of recording, transmitting and listening: sounds heard are products of pre-existing reality plus conditions of reproduction leading to specific reality, which is not the neutral transmission of a sound event or pure fabrication by technical means.

In the 1990s an online discussion held by British academics posited "radiogenic" as something having its origins in radio or designed with radio in mind specifically. This would interestingly exclude music—a major content in contemporary radio—since it does not have its origins in radio broadcasting. On the other hand, "radiogenic" means following the conventional aesthetic of radio, which is a useful notion since it challenges the idea of something that is essentially radio-like. There is also a practical component to the concept: in radio drama, a protagonist's thoughts presented as an interior monologue can be considered radiogenic, as can environmental sound effects. Furthermore, stories that people tell in their own voices might work better in radio than in television or other medium (Chignell, 2009, pp. 93–94). The listener's narratives, accompanied by environmental sounds, are given special attention while planning and thematising the content for *Evening of Sounds*, and the nationwide scope of the broadcast brings dialects from different parts of the country into dialogue with one another, adding yet another layer to programme's radiogenic content.

In the early history of terrestrial radio the radiogenic approach was represented by sound art experiments and reportages. Russian filmmaker and sound artist Dziga Vertov established a Laboratory of Hearing in 1916 in order to produce musical-literary audio montages and documentaries. Having become frustrated to challenges and impossibility in describing diverse and complex environmental sounds, Vertov changed his focus to the visual medium of film. In 1925 he returned to the topic by inventing a concept of *Radiopravda* ("Radio Truth"). Sound recording on film enabled the reproduction of environmental sounds for use as "an acoustic montage of working life." This could be broadcast as "radio-newspaper" in order to develop "a radio ear," thus advocating "an auditory bond among the international proletariat" (Lacey, 2013). Some advocates of radiogenic art were less interested in providing art for the people than accommodating people and their listening modes to the new radio art. Radio playwright Walter Bischoff speculated on this new art of listening and how radio could restore lost sensitivity to "tone colour, breath and pitch of the spoken word [for] the degraded modern urban ear" (Lacey, 2013).

Radio differed from its media predecessors with its possibility of communicating to vast groups of people simultaneously. Composer Paul Hindemith was active in promoting and executing radio reportages on German radio and in 1926 developing what has been called a "radiogenic reproduction of reality" and broadcast under the name *Disoriented Microphone*. A reporter let the microphone pick up the sounds of Frankfurt streets, railway stations and docks without script or sound effects. The series ran for several years. Different forms of reportage were considered ideal radiogenic forms because the coverage "obeyed the same laws of singularity and transience." Also, it's worth remembering that it was a sensation when wireless broadcasts came into the public domain, arousing immense public interest as a medium that enabled participation in a new sensory experience (Lacey, 2013, pp. 103–105).

When situating the broadcasting of environmental sounds in a historical context, the work of the World Soundscape Project (WSP) also deserves mention. The WSP undertook an extensive field trip across Canada and used the collected material to form the basis of series of radio compositions on sonic environments; the ten-part programme *Soundscapes of Canada* was broadcast on CBC's *Ideas* programme in 1974. Every programme was intended to be a special listening experience in and of itself, often exploring new dimensions of the radio medium. The programmes revolved around different themes such as rhythm and tempo, ambience, acoustic space, language, soundscape composition and so on, and were designed to stimulate listeners' awareness of sound and its perception (Truax, 1978/1999). In this context *Evening of Sounds* draws from the WSP legacy while remaining distinct in terms of thinking differently about the role of technology and the archives when listening to environmental sounds (see, e.g. Sterne, 2013). From a musicological or ethnomusicological point of view, musical and environmental sounds are approached as culturally specific manifestations—be they recorded or live—and approached functionally and aesthetically as part of the culture where they are performed, heard and listened to.

The role of technology in fostering relationships between people and their environments can be seen as part of the sensory turn in soundscape research, which is interwoven with technological developments and their applications in ethnographic approaches. Starting from tape recorders and video cameras in the late twentieth century, anthropological knowledge was given "an audiovisual cast," thus further challenging the language- and text-based methods of the previous decades. We make sense of the world not only through language but through our senses and their extension in the form of diverse media (Howes, 2013). The questions of music, environmental sounds and technology can be contextualised to wider notions of human-technology-world relationship. Philosopher of science and technology Don Ihde (2001) presents four modes for this relationship: embodiment relations, hermeneutic relations, alterity relations and background relations. Listening to environmental sounds broadcast live could be interpreted as embodiment relation, an augmentation of reality, but also as hermeneutic relation where the world's sonic manifestations are interpreted differently because of technology. The listeners are at the same time aware of this process and how technology alters, frames and constrains it (Kytö, 2018).

Evaluated ontologically, the sounds broadcast in *Evening of Sounds* are representations, reflections and framings of sounds and soundscapes. In this sense, the programme resembles the introduction of the modern tape recorder in the 1940s and how it was referred to as a "sound mirror": instead of implied objectivity and neutrality of a machine, it reflects an image of sound to the listener and—more importantly—it frames reality and makes us more intensely aware of it (Truax, 2001, p. 219). The concept of sound mirror actually adds yet another dimension to radiogenic broadcasting. Like a mirror, the sound recording and its airing frames the soundscape and represents this selected piece of sonic environment back to the radio listeners anew.

Evening of Sounds positions itself between conventional radio programmes such as music call-in shows, citizen's forums and special programmes. Sounds are retrieved and requests fulfilled on air, and the diverse opinions and narratives are listened to and scholarly topics of soundscape studies, ethnology and folklore are introduced. Furthermore, like music phone-ins, the programme is enabling the sharing of the soundscape-related preferences and experiences with other radio listeners. This radiogenic intersubjectivity is especially clear with environmental sounds, since they seem to evoke collective memories and emotions. In this respect, the programme content can be approached through appeals to the ontology of music and sound. Although mediated sounds carry meanings the listener has attached to them, they are still connected to place and time once heard. The process of converting sound events to sound objects enables extinct sounds to be heard again and experienced collectively on the radio.

The Requests

> According to *Act on Yleisradio oy*, The company shall be responsible for the provision of versatile and comprehensive television and radio programming with the related additional and extra services for all citizens under equal conditions. These and other content services related to public service may be provided in public communications networks nationally and regionally. In particular, it shall support democracy and everyone's opportunity to participate by providing a wide variety of information, opinions and debates as well as opportunities to interact. (Finlex, 1993).

Parallel to these responsibilities, *Evening of Sounds* offers a platform for listeners to participate in the programme and to have an impact on its content by requesting sounds via email or by telephone. It also encourages attentive listening and discussion on diverse soundscape-related themes. The criteria for selecting sounds are related to their aesthetic and narrative diversity. Also, since the third programme, the sound objects and soundscapes could be retrieved from the archives during the live show. With this possibility, the content and the structure of the programme have become more adaptive in fulfilling the requests of the callers, while remaining simultaneously unpredictable.

The sound requests are frequently related to collective and personal memories from childhood and adolescence and to soundscapes that no longer exist. These generational sound memories are especially notable and to some extent identical, when they are connected to exceptional and emotionally challenging situations, such as wartime recollections represented in autobiographical texts (Uimonen, 2016). The most wanted sounds originate from the agricultural environment. This can be partly explained by their radiogenic nature but also by the demographics of the listeners and late urbanisation of Finnish society compared to other Western European countries, taking place relatively quickly in the 1960s (Kolbe, 1999). This suggests that radio listeners have personal experience in dairy farming or fieldwork. From 2012 to 2015 the most requested sounds are as shown in Fig. 14.1.

Quite interestingly, many of these sounds are not related to a given environment per se, but to machinery needed to perform certain agricultural tasks. Also, sounds are not requested because of their functionality or the qualities they might refer to, but seemingly because of the memories and personal experiences connected to these sounds. Sounds of the separa-

1) Thresher (20 requests)	6) Separator (12)
2) Horse-powered reaping machine (16)	7) Telephone wires singing (12)
3) Steam engine (15)	8) Hand milking (11)
4) Zetor tractor (13)	9) Scythe sharpening (7)
5) Wickström engine (13)	10) Thunderstorm (6)

Fig. 14.1 Most requested sounds on *Evening of Sounds* from 2012 to 2015

tor evoked memories from the 1950s, when "holiday season in grandparents' farm began. The farmhouse offered all the joys of the summer such as making hay, taking cattle to the pasture, swimming and playing" (ÄI2T, 2012; ÄI8TS, 2015). This effect of reminiscence when a past atmosphere is brought back to listener's consciousness by a particular sound or soundscape is called *anamnesis*, whereas the concept of *phonomnesis* refers to mental activity that involves internal listening, which includes creating sound textures for composition and recalling to memory sounds linked to a given situation. In the absence of actual sound, phonomnesis can be created, for example, from reading a description of a sound, which leads to the conclusion that listening can be a totally mental and silent act (Augoyard & Torgue 2005, p. 21, p. 85). Phonomnesis is actualised when callers are describing their requests and speaking about their experience before the associated sounds are actually broadcast.

The broadcast sounds had not only onomatopoetic but also aesthetic and even patriotic qualities. The requests included vivid description of the rhythms of different tractor engines and the alleged semantic content heard in these rhythms, which unfortunately might get lost in translation. The Czechoslovakia-based manufacturer Zetor was especially popular. Its engine sound was characterised as something that can be interpreted as the protestant ethics of hard work and discipline:

> The post-war Finland was built on: it was used for ploughing the fields and rock removal, to tow away the logs from the forest in the winter, it was used to for hauling the hay and for transporting the milk churns. It [sound] represents active and hard work. (ÄI2T, 2012; ÄI3T, 2012; ÄI6T, 2014)

Radiogenic intersubjectivity was fulfilled when the programme accidentally broadcast a sound of a tractor the model of which was not precisely the one requested. The next caller kindly corrected the hosts and invited them to come to meet him and to listen to and record the engine of *his* tractor.

Meanings connected to environmental sounds transform over the years. This transcoding, reappropriating existing material for new meanings (Uimonen, 2003), also changes the acoustic community's soundscape competence. Transcoding includes names, as when proper nouns are transformed to common nouns. The name BMW, for example, has shifted from its reference to the German car to the distinctive-sounding stationary engines manufactured by Björneborg Mekaniska Werkstad in Pori, Finland. These combustion engines were run by gasoline, kerosene or heating oil and were used to give power to circular saws, shingle machines and flour mills. A stationary engine called Wickström was characterised as a seasonal sound, because the sounds of the threshers were heard in the autumn (ÄI6T, 2014; PK, 2016; MM, 2016).

> The chunking of the Wickström engine is inseparable part of my childhood and adolescence soundscape. You could hear in the autumn, since they fed power to threshers. Especially the ignition phase, when the decompression valve was used to reduce the compression. (ÄI6T, 2014)

The effect of the versatile combustion engines has been interpreted as a single sound source, which has homogenised the soundscape to a large extent. However, these sounds are considered personally meaningful in different environments, including their diverse and nostalgic contemporary use in various contexts as presented in YouTube clips. The distinctive-sounding engines also challenged the line between rural and urban sounds. The Wickström (2017) was used to power the sea and lake boats, but it could be heard also in urban environments as remembered in another request: "[O]n Saturdays, when residents of Kotka (Finnish coastal town) sailed to the islands, air was filled with this beautiful music" (ÄI2TS, 2012). The motors are pleasant to listen to; unlike the ideally steady-running contemporary engines, they sound irregular and unpredictable, thus catching and keeping the listener's attention.

Another often-requested sound, now extinct, was called "telephone wires singing." The sound is a result of metal telephone wires strung between telephone poles, which made a distinctive sound on cold and quiet winter nights. The metallic wires were attached to petticoat insulators and in mildly windy weather conditions the wires started to vibrate. The transmission of the acoustic energy through a medium could be listened to more closely by placing the ear against the wooden telephone pole. The listener made this conscious effort at night, which seemingly further enhanced the experience.

> Sound of the telephone wires was part of my childhood in the 60s [...] It reminds me of full moon and cold weather [...] Their sound was empowering and soothing at the same time. Sometimes the howling and singing of the wire attached to the wall kept me awake. When the time was right I liked to walk on a snow crust[2] on a full moon from one pole to another and press my ear to the wooden pole and listen to it sing! (ÄI7T, 2014)

A recollection also describes how transformation of sounds and generational soundscape competence are intertwined with technological improvements, digitisation and the transforming mediascape. Contemporary coated wires don't make sounds like the metallic ones, and gradually the mobile phones replaced the landline telephones thus silencing the wires altogether. A popular culture artefact might explain why this sound became so well liked: The recognisable sound of the metal wires between the poles was turned into a Finnish schlager/schottische in 1947 and again in 1964 titled "Puhelinlangat laulaa" (Telephone wires are singing). These cultural meanings were also referred to by the writer, who suspects that the younger generation might not know the meaning of the phrase. And for yet another layer, a radio programme of the 1970s broadcasting song requests was also named after the singing telephone wire phenomenon.

Feelings of Safety and Fear

Evening of Sounds is a thematised and pre-planned radio programme. Selected sounds and soundscapes retrieved from the archives are composed in accordance with the requests. These preliminary preparations enable the broadcasting of requested sound collages which frequently step outside the realm of realism, such as the "sound of approaching train and swarm of bees together." Disappearing and extinct sounds prove to be especially meaningful. As noted, the requests are very often related to memories and nostalgic feelings of safety (see Järviluoma, 2006), such as soundscapes reflecting an implicit wish to return to the endless warm and calm summers of childhood "when people were happy." One particular wish for a sound collage was described as follows:

> A sound memory from the summers of my childhood and adolescence consists of a combination of lapping of waves, sounds of the old motor boats, the landing and departing of the inland waterway boat with its whistles, and screams of joy from the nearby beach. Also, the collage includes the sound of a light tourist aircraft, since you did not hear this on bad weather. (ÄI4T, 2013)

In a geographical context, the climate conditions determining seasonal sounds are worth noting. A request for childhood summer soundscapes is more likely to be made in a country with long, cold and dark winters combined with desperate anticipation of warm and bright summer days. Similar summer feelings were described in other requests related to childhood and feelings of safety such as "starling singing in the break of dawn." As one listener shares: "When I was a child in the 1950s there was a starling's birdbox in the yard. In summertime, I used to sleep in the granary and woke up to the singing of the birds" (ÄI4T, 2013). Apart from dawn chorus soundscapes, the countryside is represented by the animal sounds such as cows mooing, sound of a cowbell, cats, dogs and sheep. Sound of water and rain were requested, including rapids, waves and especially a soothing rainfall on a shingle roof.

A broadcast sound object is usually related to the surrounding soundscape and it should not be detached from it. On the other hand, an individual sound can trigger a nostalgic memory, which turns out to be a multimodal one: visual, olfactory, tactile and gustatory. A sound of a slap shot brings back memories of an "ice hockey rink, vaguely audible music, blue-black evening, smell of a locker room, ever-lasting struggle with the skate laces, and warm juice, which was sold next to the rink" (ÄI3TS, 2012). Environmental sounds or an individual sound event can be listened to attentively and intellectually, but at the same time, like music, "most listeners have little rational control over the way it influences them" (McClary 1991, p. 151).

Not only pleasant sounds, but also undesired—or even sometimes highly unpleasant—sounds such as dentist drills are connected to multimodal and specifically tactile experiences (ÄI5TS, 2013). These affect-related and sound-triggered memories are well remembered in bodily ways. Some recalled a safe ride on a two-wheeled vehicle when someone trustworthy, such as a father, was steering the bicycle or a moped. The sound of a gravel road under a bicycle wheel or of a particular moped's engine were attached to these moments experienced 50 years ago (ÄI2T, 2012; ÄI2TS, 2012).

Tellingly, unfamiliar sounds were also frequently requested. These curiosities are not necessarily tied to any particular place, although the requests are usually reasoned in detail. ("My child was learning the sounds of animals and wondering if the long neck of the giraffe is making a low or a high-pitched sound") (ÄI3T, 2012). These seemingly innocent requests might lead to undesired results. After broadcasting a sound of the howling pack of wolves, *Evening of Sounds* received feedback about nervous household cats and a frightened child (ÄI1TS, 2012; ÄI5T, 2013).

In the context of feelings of safety and fear, Karin Bijsterveld's concept of *auditory topoi* is fruitful, used for extending the categories of sound developed by Schafer and the World Soundscape Project for use in analysing the dramatisation of sounds in radio plays, television and films (Bijsterveld, 2013, p. 15). Of the four *auditory topoi*—intrusive sound, sensational sound, comforting sound and sinister sound—particularly interesting ones are the intrusive and comforting sounds. The former invades or threatens the vulnerable of fragile existence, whereas the latter sound refers to a single source of sound, often heard at a certain distance and from an unspecified direction (Bijsterveld, 2013, p. 19).

Some requests simply cannot be fulfilled, such as the sound of an author's favourite typewriter, or "total silence." The former can be perhaps interpreted as an encouragement to document a sound seemingly valuable to the listener and also to national literary canonisation. Requests on silence and quietness included descriptions of sonic environments and thus can be labelled as subjective silences (ÄI1T, 2012; ÄI4T, 2013). This sometimes-contradictory discourse was crystallised by historian Peter Englund, who stated that "[s]ilence does not exist and it has a long history." The history of silence is intertwined with its opposite, which is the history of sound (Englund, 2003, p. 14). This analytical approach understands silence as an anomaly, but also by definition as something that exists in relation to something which is audible or perceived by other senses. Englund's notions are parallel to themes discussed in *Evening of Sounds*, especially when related to transforming soundscapes. While studying or discussing sound, a historical context should not go unnoticed: our soundscape has very little in common with the one 200 years ago, not the least because sounds were perceived and interpreted differently in those days than we do now (Englund, 2003, p. 28). This is also one of the leading principles of the programme: not only to broadcast meaningful historical sounds to radio listeners, but to understand sounds as something that has different meanings for different people in different times.

Auditory topoi are detectable not only in the representations of urban soundscapes in textual or other media formats (Bijsterveld, 2013, p. 20) but also on memories related to personal and collective experiences of media. The popularity of communally understood sounds is partly explained by the unity of the Finnish media culture constructed by the limited number of radio and television channels, until the mid-1980s' deregulation introduced commercial and local radio stations.

Media Memories

In *Evening of Sounds* several requests have been related to the FBC's programme content or auditory signals used in radio broadcasting. Broadcast call signs, time signals and voices of the familiar radio announcers have become part of the media memory due to the fact that this content was repetitive and heard on a daily basis. Since the founding of the FBC in 1926, they have constructed an identifiable soundscape of the radio—or a channel sound, as it would be called in contemporary terms referring to the overall sonic character of a station (Uimonen, 2011, p. 31). Radio-signal requests are an assorted sample of FBC as a constructor of a nation's media memory, including a rooster crow broadcast in the mornings from the 1930s to the 1960s (Matilainen, 2015).

Public service responsibilities of the FBC included announcing the exact time to radio listeners. A small table gong used in the 1950s was substituted with a series of six individual electronic beeps with the gong announcing the last one until 1972, when it was discarded as old-fashioned. In 1975, the time signal was automated and broadcast on the hour without the specific announcement. The six "pips" are identical to the Greenwich time signal broadcast by the BBC, which was first heard in 1924. At FBC the time signal was not allowed to disturb programme contents of artistic value such as a symphony orchestra concert (BBC, 2017; Ilmonen & Tommila, 1996, p. 142). The practice revealed the existing music genre hierarchies of the public service radio and that all music genres were not treated equal: those who were home-taping popular music to their compact cassettes soon learned that the time signals ruined their recording sessions (Kilpiö, Kurkela & Uimonen, 2015).

In Europe, the use of broadcast call signs and distinctive signals was established during the early years of public service broadcast stations. Signals were used to differentiate broadcasters from each other, since programmes could be listened to transnationally. As part of the public service, the FBC's printed publications disseminated programme information of the other European radio channels' content. As a rule, the easily recognisable melodies were selected as call signals. A chart from the year 1937 includes a brief description and notations of pre-WWII signals in Europe. It reveals that melodies' broadcast on 47 stations consists of opera melodies or folk songs excluding Northern Italy and, Vilnius, then part of Poland, which broadcast nightingale song and cuckoo, respectively (Ilmonen & Tommila, 1996, p. 143; Lyytinen, Tommila, & Vihavainen, 1996, p. 59).

Sonic construction of the pre-war nation and the analogue technological advancements were represented in FBC's interval signal. Its melody was based on a pentatonic folk tune from the eighteenth century and arranged by music scholar and later professor of musicology Mr. Armas Otto Väisänen. The first version was recorded with a Finnish folk instrument *kantele*, which is a plucked five-string instrument belonging to the zither family and is called "Finnish National Instrument" partly due its significant role in the Finnish national epic *Kalevala*. The initial version of the interval signal of 1930 was replaced the following year with a mechanical five-string soundbox, which was specifically constructed to play the tune. The mechanical soundbox was substituted with the electronic one in 1960 and in 1978 the interval signal was computerised (Ilmonen & Tommila, 1996, p. 143; Lindfors, 2008). Familiar voices of the FBC's announcers such as Kaisu Puuska-Joki (1907–1996), Carl-Erik Creutz (1911–2000) and Pentti Fagerholm (1935–2015) were considered memorable by many requests. Also, the announcements on Marine Weather or "Announcements for Seafarers" represented safety and feelings of continuity for many radio listeners. Repetitive reading and the ritual nature of the semantic contents resembled the act of meditative listening—or at least the semantic content was of minor importance to some of the listeners. The affect is composed of a peaceful rhythm and clear articulation of words in the message, which is usually highly predictable. At the same time the announcement represents the somewhat official and bureaucratic but also safe, fatherly, loving and caring Finnish welfare state. One caller remembered listening to a Finnish reading of "Marine Weather" to cure her homesickness while studying in the German Democratic Republic (ÄI4TS, 2012).

Discontinued information briefings of the FBC are often requested, too. Their nostalgic ambiance is enhanced by the official and literal parlance of the public service station. A briefing called Stereotesti ("Stereo Test," 1972–1990) represented the enlightened media policy of the FBC, broadcast for radio listeners with stereo equipment to make sure that they have plugged in their gear properly (including a series of analogue metronome clicks used to adjust loudspeaker balance) (ÄI8TS, 2015; Aunila, 2006; Lindfors, 2008). Also, samples of individual radio and television plays are often requested, including their diegetic and non-diegetic sounds, jingles, children's rhymes and sayings. These requests reflect the memories of several generations and the transformation of the media they have lived through. Phrases and sayings from the discontinued programmes have continued their existence in medialore and become part of today's social interactions when media contents have been adapted to an oral tradition (Julkunen, 1989, p. 49).

Interestingly, callers have also requested radio-specific sounds from the predigital broadcasting era, such as the unwanted noises associated with improperly tuned stations on the dial and signal interference, described as the "buzzing of radiowaves and chaos of sounds" and "sounds of dialling the station such as gurgling, whining, beeping, rattling and crackling" (ÄI1T, 2012; ÄI6T, 2014). Digital technology sounds have also been requested, especially important given how fast these technologies change and become obsolete. And so callers would ask for the sounds of past computer programs, mobile phone ringtones and data transmission like FAX and dial-up Internet modems (ÄIT8T, 2015), which clearly underlines the importance of documenting our continuously transforming everyday soundscapes along with their contextual information.

Broadcasting the Archived Material

Evening of Sounds made use of present-day soundscape recordings as programme content. They were provided by research projects carried out by the Finnish Society for Acoustic Ecology including *One Hundred Finnish Soundscapes* and *Transforming Finnish Soundscapes* (OHFS, 2006; TFS, 2017). However, a programme consisting primarily of sounds of the past necessitates not only a comprehensive archive of sounds but also access to these archives. The FBC sound archive was founded at the same time as the FBC itself in 1926. The archive was originally designed and created for the needs of radio drama. The following year a total of 33 radio plays were broadcast. The imported sound effects were supplemented by the FBC's own archive, which was accumulated in accordance with the principles of locality and self-sufficiency (Lyytinen, Tommila & Vihavainen, 1996, p. 99). The Foley effects for the radio plays were performed manually and in real time in radio studios, with different mechanical devices constructed for this particular purpose. Sound effects were collected first on lacquer/instantaneous discs, with sound inscribed onto aluminium plates covered by lacquer (Ilmonen & Tommila, 1996, p. 51; p. 143; Lindfors, 2006). Later, the Tehosto sound archive would be founded in 1957, also with Radioteatteri (Radio Theatre) in mind.

The practice of collecting sound in the field increased dramatically when schooling for sound engineers was established. During the first year, almost 1800 different sound samples were collected with "scientific accuracy": a motorcar was documented from inside and outside, from different distances, while ignited and accelerating. Field recording was further enhanced by evolving technologies for sound recording. The professional use of magnetic tape in the 1950s helped the process of collecting environmental sounds and made their end use more diverse compared to the relatively cumbersome lacquer discs. Also, the sound quality of magnetic tape was superior to lacquer discs, which came with a lot of extraneous and undesired noise. The early portable, spring-driven Maihak tape recorders enabled the mobility of reporters in the field, but were replaced in the 1960s with the more reliable Nagra recorders that became the professional standard. Along with accounting for these technological changes, training programmes for sound engineers also began to take notice of social changes, leading to an increase in recording contemporary sounds of the 1950s and the 1960s while also gathering material associated with times past. Currently there are over 80,000 sound effects in The FBC's sound effect collection (Gronow, 2010, pp. 181–190; Ilmonen & Tommila, 1996, p. 144; Kuljuntausta, 2005, p. 126, p. 130–131).

For *Evening of Sounds* the archive was put to use in broadcasting sound requests such as: announcements at horse races, old diesel-engine train toilets, drills in the Finnish army, biting and chewing an apple, Maxim machine guns, cutting Styrofoam, the keypad of a Commodore 64, split-flap displays, speaking clocks/Neiti Aika (Miss Time), and Pirunviulu (the Devil's Fiddle, which is a piece of string pressed against the window and rubbed with a piece of resin to produce a piercing sound).

The original purpose of FBC's sound archive is in fertile contradiction with the fundamental idea behind *Evening of Sounds*. Archived sounds or soundscapes were not meant to be listened to attentively or broadcast as such, but instead to support radio plays and other narratives in radio content. Sound events transformed into sound objects become floating signifiers. They become signifiers when a radio play is constructed from various sound objects in order to support the narrative. However, a particular sound or a soundscape—while listened to attentively—is already telling a story or bringing memories back to listeners. Furthermore, the FBC is able to prove its public service responsibilities for listeners with various accessibility issues, such as presenting the sound of a purring cat requested by an allergic radio listener, or distinctive sounds of bathing in the sauna, a request made as a gift from a wife to her husband who had to remove his hearing aid before bathing (ÄI4TS, 2013; ÄI8T, 2015).

The uses and limitations of archived material should not go unnoticed. Although archives enable the experience of pastness, they should not be considered something that offers objective evidence upon which history can be built: archives are systems that govern what can be said about the past, thus representing a particular power structure in the writing of history (Baron, 2013). This also holds true with archival sound recording, especially the motives driving the recordists. Making the archive flexible for alternative uses has been achieved, in part, by including the recording site's contextual metadata with the actual recordings. From an ethnomusicological point of view, the issue of contextual data is of utmost importance, especially when recordings are not only archived but released for a wider audience.

Archival policies, especially the reuse of recordings, are intertwined with issues of ownership, colonialism, globalisation and the politics of music dissemination, as pointed out by ethnomusicologists and sound anthropologists when discussing the recording and distribution practices of so-called world music (Feld, 2000). The practice of providing adequate information on liner notes about recording sites and musicians, as well as the motives of the recordists, would contribute ethical archival practices and help future scholars to contextualise historical recordings, no less important for soundscape material. The meanings embedded in any given soundscape depend entirely on the context in which they can be heard, experienced and interpreted spatially, temporally and socially. Soundscapes are the sonic environments of the listeners, without whom they would not exist (Kytö, Remy & Uimonen, 2012).

Archives, then, can be said to create the past more than simply preserving it. Instead of history, they contain many diverse histories open to multiple interpretations and experiences. The simultaneous tangibility and ambiguity of sound makes this material even more unruly than text. Sound seems to give the impression that it is closer to the past it represents. Jaimie Baron writes about "unruliness of indexical signs," referring to sounds that, when framed in a certain way, can create perceived authenticity among radio listeners; the rarity of a particular sound can imbue its recording with truth value, be it true or not (Baron, 2013, p. 3, p. 6), which might explain part of *Evening of Sound*'s popularity: seldom heard sounds have an aura of authenticity that is lost in the soundscape we are surrounded with every day.

Conclusion

Radiogenic experiences are framed by artistic, political and technological factors. Vertov's point of departure was political in his attempt at creating an auditory bond for the proletariat. Granting broadcasting licences for terrestrial radio is part of the national decision-making process, which is inherently political. The result in most countries is usually a radioscape composed of European public and American-style commercial radio. The *Evening of Sounds* live programme is both restricted and enabled by the content requirements of Finnish public service radio and its economical, technological and intellectual resources, including the nationwide range of the transmitter, the programme's place in the station's schedule and the archives provided by the FBC. Privately owned radio stations, predominantly oriented around economic concerns, would have found it challenging to broadcast any major content other than music in the relatively limited market of Finland. Soundscape-related programmes have been rare in Finland, with the exception of a couple of 1990s' commercial stations, which broadcast the sound of birdsong during the night-time. Tellingly, this practice was not the result of an interest in soundscape-oriented content, but was introduced for economic reasons: to reduce the overall percentage of the music content in the context of total air time, which resulted in higher levels of remuneration for the station (Uimonen, 2011, p. 114; p. 149).

Evening of Sounds amalgamates different types of radio programmes. The major content of the programmes are the memories and feelings of the listeners, and the sharing of the sounds that prompt these memories and feelings with the broader radio community. Together with media-related memories, the programme also generates sonic phenomena for potential future audiences. Formerly everyday sounds have become affective in their capacity to trigger emotional responses and thus construct and maintain acoustic and electroacoustic communities. The triangular relationship between listeners, hosts and broadcast sounds is reinforced by the contextual information and experiences of the callers. Alongside them, the phonomnetic requests and oral descriptions form an integral part of the programme framing these tangible yet unruly sounds.

It is worth asking whether environmental sounds should be evaluated as music-like phenomena or "organised sounds", exposing them to aesthetic criteria applied to music and diverse music genres. The topic can be approached with the help of radio as a medium for disseminating sounds, and how this dissemination turns quotidian sound objects into something special. New meanings are attached to music when it is transported to new contexts, which can happen to any environmental sound or radio signal when listened to attentively. In a world of mediated sound and listening we should not restrict our discussion to contemplative listening rooted in the tradition of Western classical music, or to consider the world like a compositional work as suggested by early soundscape scholars (see Sterne, 2013, p. 191). Music scholar Anahid Kassabian's (2013, p. 18) notions of ubiquitous listening and ubiquitous music refer to listening as a simultaneous or secondary activity, with implications in many directions such as modulating our attentional capacities, tuning our affective relationships to categories of identity and conditioning our participation in fields of subjectivity. Like ubiquitous music, everyday and allegedly extinct sounds can be affective, helping to construct the listener's identity through oral narratives of remembrance.

Radiogenic listening to environmental sound content is somewhat scarce. It might become a unique experience for the listener, unlike recurrently heard individual tunes or songs. The uniqueness of specific sound recordings is further underlined by the fact that the generational experience of them usually goes without comment, unlike musical performances presented and discussed at length in the media. This difference harkens back to my earlier suggestion that a show like *Evening of Sounds* runs parallel to music call-in programmes as they both function to share both sonic material for their own sake and the experiences that listeners bring to them. This is enabled by the fellow listeners' soundscape competence, referring to the ability to interpret sounds of a given cultural background and to value sounds' social and perhaps aesthetic implications. With these parallels in mind, the concept of "radiogenic sound" needs to be developed further. In emphasising the ethereal connection between a sound and its capacity to sound good over the radio, current definitions fail to take into consideration the sociocultural background of the radio listeners, which sets the frame for their soundscape competence and the affective qualities of sound. The idea of radiogenic sound should also be made relational to the contemporary media environment, since innumerable websites, streaming services and podcasts are now offering experimental content for radiogenic listening (see, e.g. Virtual Barber Shop VBS, 2017).

Clearly, like any piece of music, soundscapes should be studied in their social, cultural and historical contexts. However, this does not automatically exclude the fact that other sound cultures could inform sonic evaluation of any given soundscape recording. Thus *Evening of Sounds* also raises questions of nationality and homogenic culture. In 2016, when the radio programme was presented and discussed at Crossroads in Cultural Studies, English-born Australian sound artist Dr. Nigel Helyer pointed out that collective meanings attached to given sounds would be unlikely in his home country. Also, sound artist and scholar Dr. Yeung Yang from Hong Kong stated that a public service radio in her home country airing nostalgic sounds from the past would be problematic and might be interpreted as a hegemonic voice of the powers-that-be. This suggests that sounds interpreted and labelled as "unique" and especially "national" should be critically re-evaluated, as with any cultural products that bear such an epithet.

In the *Evening of Sounds* programme one could find echoes of banal nationalism referring to less visible practices through which the nation is being constructed (Billig, 1995; Kytö 2018). Like any cultural artefact, radiogenic sounds can be remembered and placed within a national narrative, regardless of their transnational origin. Soundscapes of the sometimes mythical post-war reconstruction era were represented by the sound of the Czechoslovakian-made tractor. These sounds are nevertheless products and representatives of their own era, and they vanish when they are no longer functional or aesthetically pleasing to their listeners. Like any national narratives, sound memories are the result of selection, and this selection inevitably includes some sound events and excludes others. *Evening of Sounds*, at least, offered the opportunity to democratise the process of selection on a national level, allowing its listeners to participate in the construction, reconstruction and deconstruction of their sonic identities.

Notes

1. I would like to thank Dr. Meri Kytö for the many discussions on the subject and access to his article manuscript on Mediated Soundscapes. The article in based on our keynote lecture "Radiogenic Sound Cultures in Call-in Programmes and the Question of National Narrative," at the conference *Music, Sound, Radio: Theorizing Music Radio*, 25 May 2017 in Copenhagen and an article published in Finnish by Uimonen and Mikkola in Uimonen, Kytö and Ruohonen (eds. 2017) *Muuttuvat suomalaiset äänimaisemat.* The article received funding from ERC AdG 2015 project 694,893 Sensory Transformations and Transgenerational Environmental Relationships in Europe, 2016–2020.
2. In the springtime a snow crust freezes to the point that it can support a person walking on it.

References

Radio Programmes

ÄI1T (2012). 1st Äänien ilta radio programme 17.1.2012 (requested).
ÄI1TS (2012). 1st Äänien ilta radio programme 17.1.2012 (requested and broadcast).
ÄI2T (2012). 2nd Äänien ilta radio programme 14.2.2012.
ÄI2S (2012). 2nd Äänien ilta radio programme 14.2.2012 (broadcast).
ÄI2TS (2012). 2nd Äänien ilta radio programme 14.2.2012.
ÄI3T (2012). 3rd Äänien ilta radio programme 13.3.2012.
ÄI3TS (2013). 4th Äänien ilta radio programme 10.3.2013.
ÄI4T (2013). 4th Äänien ilta radio programme 10.3.2013.
ÄI4TS (2013). 4th Äänien ilta radio programme 10.3.2013.
ÄI5T (2013). 5th Äänien ilta radio programme 13.10.2013.
ÄI5TS (2013). 5th Äänien ilta (Evening of Sounds) radio programme 13.10.2013.
ÄI6T (2014). 6th Äänien ilta radio programme 16.3.2014.
ÄI7T (2014). 7th Äänien ilta radio programme 12.10.2014.
ÄI7TS (2014). 7th Äänien ilta radio programme 12.10.2014.
ÄI8TS (2015). 8th Äänien ilta radio programme 8.3.2015.
ÄI8T (2015). 8th Äänien ilta radio programme 8.3.2015.

Literature

Augoyard, J.-F., & Torgue, H. (2005). *Sonic Experience. A Guide to Everyday Sounds* (A. McCartney, & D. Paquette, Trans.). Ithaca, NY: McGill-Queen's University Press.

Aunila, S. (2006). Radion stereotesti. *Yle. Elävä Arkisto*. Retrieved from http://yle.fi/aihe/artikkeli/2006/12/01/radion-stereotesti

Baron, J. (2013). *The Archive Effect*. New York: Routledge.

BBC. (2017). *BBC Magazine Monitor*. The Comforting Tone of the Hourly Radio Pips. Retrieved from http://www.bbc.com/news/blogs-magazine-monitor-26052087

Bijsterveld, K. (2013). Introduction. In K. Bijsterveld (Ed.), *Soundscapes of the Urban Past. Staged Sounds as Mediated Cultural Heritage*. Wetzlar: Humboldt-Universität zu Berlin.

Billig, M. (1995). *Banal nationalism*. London: Sage.

Chignell, H. (2009). *Key Concepts in Radio Studies*. London: Sage.

Chion, M. (1994). *Audio-Vision. Sound on Screen*. New York: Columbia University Press.

Englund, P. (2003). *Hiljaisuuden historia*. Helsinki, Finland: WSOY.

Feld, S. (2000). A Sweet Lullaby for World Music. *Public Culture, 12*(1), 145–171.

Finlex. (1993). Laki Yleisradio Oy:sta. *Finlex*. Retrieved from http://www.finlex.fi/fi/laki/ajantasa/1993/19931380#L3P7

Gronow, P. (2010). *Kahdeksas taide. Suomalaisen radioilmaisun historia 1923–1970*. Helsinki, Finland: Avain.

Hannula, M. (2016). *Personal Communication*. Email from Producer Minna Hannula to Jukka Mikkola, Yleisradio.

Howes, D. (2013). *The Expanding Field of Sensory Studies*. Retrieved from http://www.sensorystudies.org/sensorial-investigations/the-expanding-field-of-sensory-studies/

Ihde, D. (2001). *Bodies in Technology*. Minnesota, USA: University of Minnesota Press.

Ilmonen, K., & Tommila, P. (1996). *Yleisradio historia. 3. osa, 1926–1996*. Helsinki, Finland: YLE.

Järviluoma, H. (2006). Turvallisuuden tunne, äänimaisema ja eletty tila lapsuudenmuistoissa. In H. Järviluoma, A. Koivumäki, M. Kytö, & H. Uimonen (Eds.), *Sata suomalaista äänimaisemaa* (pp. 23–40). Helsinki, Finland: SKS.

Julkunen, E. (1989). Lasten ja nuorten joukkotiedotusperinne. In J. Pöysä (Ed.), *Betoni kukkii. Kirjoituksia nykyperinteestä* (pp. 49–63). Helsinki, Finland: SKS.

Kassabian, A. (2013). *Ubiquitous listening. Affect, attention, and distributed subjectivity*. Berkeley and Los Angeles: University of California Press.

Kilpiö, K., Kurkela, V., & Uimonen, H. (2015). *Koko kansan kasetti. C-kasetin käyttö ja kuuntelu Suomessa*. Helsinki, Finland: SKS.

Kolbe, L. (1999). Suomalainen kaupunki. Historian hierarkioita ja modernia lähiöelämää. In M. Löytönen & L. Kolbe (Eds.), *Suomi. Maa, kansa, kulttuurit* (pp. 156–170). Helsinki, Finland: SKS.

Kuljuntausta, P. (2005). Uusi uljas äänimaailma. Ympäristöäänet, äänimaisemat ja radiotehosteet virikkeiden antajana 1950–60-lukujen suomalaisessa akustisessa taiteessa. In O. Ampuja & K. Kilpiö (Eds.), *Kuultava menneisyys. Suomalaista äänimaiseman historiaa* (pp. 122–157). Turku, Finland: Turun historiallinen yhdistys ry.

Kytö, M. (2018). Mediated Soundscapes. Representations of the National in the Soundscape Call-in Programme *Äänien ilta*. In S. Kaargaard Nielsen, M. Krogh, I. Have, & M. Michelsen (Eds.), *Music Radio: Building Communities, Mediating Genres*. London: Bloomsbury.

Kytö, M., Remy, N., & Uimonen, H. (2012). *European Acoustic Heritage*. University of Applied Sciences and Cresson.

Lacey, K. (2013). *Listening Publics. The Politics and Listening in the Media Age*. Cambridge, UK: Polity Press.

Lindfors, J. (2006). Ylen ensimmäinen äänityslaite oli mullistava uutuus. *Yle. Elävä Arkisto*. Retrieved from http://yle.fi/aihe/artikkeli/2006/09/08/ylen-ensimmainen-aanityslaite-oli-mullistava-uutuus

Lindfors, J. (2008). Radion väliaikamerkki. *Yle. Elävä Arkisto*. Retrieved from http://yle.fi/aihe/artikkeli/2008/07/04/radion-valiaikamerkki

Lyytinen, E., Tommila, P., & Vihavainen, T. (1996). *Yleisradio historia. 1. osa, 1926–1949*. Helsinki, Finland: YLE.

Matilainen, V. (2015). Aamukukot kiekuivat päivän radio-ohjelman alkaneeksi. *Yle. Elävä Arkisto*. Retrieved from http://yle.fi/aihe/artikkeli/2015/07/09/aamukukot-kiekuivat-paivan-radio-ohjelman-alkaneeksi

McClary, S. (1991). *Feminine Endings. Music, Gender and Sexuality*. Minnesota, MN: University of Minnesota Press.

MM. (2016). Maamoottori. Retrieved from *Wikipedia* https://fi.wikipedia.org/wiki/Maamoottori

OHFS. (2006). One Hundred Finnish Soundscapes. *Sata suomalaista äänimaisemaa*. Helsinki, Finland: Suomalaisen Kirjallisuuden Seura.

PK. (2016). Porin konepaja. Retrieved from *Wikipedia* https://fi.wikipedia.org/wiki/Porin_Konepaja

Sterne, J. (2013). Soundscape, Landscape, Escape. In K. Bijsterveld (Ed.), *Soundscapes of the Urban Past. Staged Sounds as Mediated Cultural Heritage*. Bielefeld, Germany: Transcript.

TFS. (2017). Transforming Finnish Soundscapes. Retrieved from *Muuttuvat suomalaiset äänimaisemat* http://www.aanimaisemat.fi

Truax, B. (1978/1999). Soundscapes of Canada. *Handbook for Acoustic Ecology*. Retrieved from http://www.sfu.ca/~truax/canada.html

Truax, B. (2001). *Acoustic Communication*. Westport, CT: Ablex.

Uimonen, H. (2003). Soundscape Studies and Auditory Cognition. In A.-V. Kärjä (Ed.), *Etnomusikologian vuosikirja 14* (pp. 21–40). Helsinki, Finland: Suomen etnomusikologinen seura.

Uimonen, H. (2011). *Radiomusiikin radiomuutos. Kaupallisten radiokanavien musiikki 1985–2005*. Tampere, Finland: Tampere University Press.

Uimonen, H. (2016). Muuttuvat suomalaiset äänimaisema ja kuulumisen politiikka. In H. Järviluoma & U. Piela (Eds.), *Kalevalaseuran vuosikirja* (pp. 15–33). Helsinki, Finland: SKS.

Uimonen, H., & Mikkola, J. (2017). Tahtoisin kuulla lehmän käsinlypsyä sinkkiämpäriin. Radio Suomen Äänien ilta radiogeenisena ohjelmana. In H. Uimonen, M. Kytö, & K. Ruohonen (Eds.), *Muuttuvat suomalaiset äänimaisemat* (pp. 23–52). Tampere, Finland: TUP.

VBS. (2017). Virtual Barbes Shop. Retrieved from *YouTube*. https://www.youtube.com/watch?v=IUDTlvagjJA

Wickström. (2017). Wickström. *YouTube*. Retrieved from https://www.youtube.com/watch?v=e8yeO-btN0I

Index[1]

[1] Note: Page numbers followed by 'n' refer to notes.

M. Droumeva, R. Jordan (eds.), *Sound, Media, Ecology*, Palgrave Studies in Audio-Visual Culture,
https://doi.org/10.1007/978-3-030-16569-7

www.ingramcontent.com/pod-product-compliance
Lightning Source LLC
Chambersburg PA
CBHW041441230526
45355CB00063BB/479
9783030165680